```
FREDERICK DOUGLASS S NARRATIVE
 OF THE LIFE OF FREDERICK DOUG
LASS


BLOOM, HAROLD


E449.D7493F74 1988
                                SC
```

Modern Critical Interpretations

Frederick Douglass's
Narrative of the Life of Frederick Douglass

Modern Critical Interpretations

The Oresteia
Beowulf
The General Prologue to The Canterbury Tales
The Pardoner's Tale
The Knight's Tale
The Divine Comedy
Exodus
Genesis
The Gospels
The Iliad
The Book of Job
Volpone
Doctor Faustus
The Revelation of St. John the Divine
The Song of Songs
Oedipus Rex
The Aeneid
The Duchess of Malfi
Antony and Cleopatra
As You Like It
Coriolanus
Hamlet
Henry IV, Part I
Henry IV, Part II
Henry V
Julius Caesar
King Lear
Macbeth
Measure for Measure
The Merchant of Venice
A Midsummer Night's Dream
Much Ado About Nothing
Othello
Richard II
Richard III
The Sonnets
Taming of the Shrew
The Tempest
Twelfth Night
The Winter's Tale
Emma
Mansfield Park
Pride and Prejudice
The Life of Samuel Johnson
Moll Flanders
Robinson Crusoe
Tom Jones
The Beggar's Opera
Gray's Elegy
Paradise Lost
The Rape of the Lock
Tristram Shandy
Gulliver's Travels

Evelina
The Marriage of Heaven and Hell
Songs of Innocence and Experience
Jane Eyre
Wuthering Heights
Don Juan
The Rime of the Ancient Mariner
Bleak House
David Copperfield
Hard Times
A Tale of Two Cities
Middlemarch
The Mill on the Floss
Jude the Obscure
The Mayor of Casterbridge
The Return of the Native
Tess of the D'Urbervilles
The Odes of Keats
Frankenstein
Vanity Fair
Barchester Towers
The Prelude
The Red Badge of Courage
The Scarlet Letter
The Ambassadors
Daisy Miller, The Turn of the Screw, and Other Tales
The Portrait of a Lady
Billy Budd, Benito Cereno, Bartleby the Scrivener, and Other Tales
Moby-Dick
The Tales of Poe
Walden
Adventures of Huckleberry Finn
The Life of Frederick Douglass
Heart of Darkness
Lord Jim
Nostromo
A Passage to India
Dubliners
A Portrait of the Artist as a Young Man
Ulysses
Kim
The Rainbow
Sons and Lovers
Women in Love
1984
Major Barbara

Man and Superman
Pygmalion
St. Joan
The Playboy of the Western World
The Importance of Being Earnest
Mrs. Dalloway
To the Lighthouse
My Antonia
An American Tragedy
Murder in the Cathedral
The Waste Land
Absalom, Absalom!
Light in August
Sanctuary
The Sound and the Fury
The Great Gatsby
A Farewell to Arms
The Sun Also Rises
Arrowsmith
Lolita
The Iceman Cometh
Long Day's Journey Into Night
The Grapes of Wrath
Miss Lonelyhearts
The Glass Menagerie
A Streetcar Named Desire
Their Eyes Were Watching God
Native Son
Waiting for Godot
Herzog
All My Sons
Death of a Salesman
Gravity's Rainbow
All the King's Men
The Left Hand of Darkness
The Brothers Karamazov
Crime and Punishment
Madame Bovary
The Interpretation of Dreams
The Castle
The Metamorphosis
The Trial
Man's Fate
The Magic Mountain
Montaigne's Essays
Remembrance of Things Past
The Red and the Black
Anna Karenina
War and Peace

These and other titles in preparation

Modern Critical Interpretations

Frederick Douglass's Narrative of the Life of Frederick Douglass

Edited and with an introduction by
Harold Bloom
Sterling Professor of the Humanities
Yale University

Chelsea House Publishers ◊ *1988*
NEW YORK ◊ NEW HAVEN ◊ PHILADELPHIA

© 1988 by Chelsea House Publishers, a division
of Chelsea House Educational Communications, Inc.,
  345 Whitney Avenue, New Haven, CT 06511
  95 Madison Avenue, New York, NY 10016
  5014 West Chester Pike, Edgemont, PA 19028

Introduction © 1988 by Harold Bloom

All rights reserved. No part of this publication may be
reproduced or transmitted in any form or by any means without
the written permission of the publisher.

Printed and bound in the United States of America

10 9 8 7 6 5 4 3 2 1

∞The paper used in this publication meets the minimum
requirements of the American National Standard for
Permanence of Paper for Printed Library Materials,
Z39.48–1984.

Library of Congress Cataloging-in-Publication Data
Fredrick Douglass's Narrative of the life of Frederick Douglass.
   (Modern critical interpretations)
   Bibliography: p.
   Includes index.
   1. Douglass, Frederick, 1817?–1895. Narrative of the life of
Frederick Douglass, an American slave. 2. Abolitionists—
United States—Biography—History and criticism. 3. Afro-
Americans—Biography—History and criticism. 4. Slaves—
United States—Biography—History and criticism   I. Bloom,
Harold.   II. Series.
E449.D7493F74 1988   973.8′092′4 [B]   87-13723
ISBN 1-55546-014-3 (alk. paper)

Contents

Editor's Note / vii

Introduction / 1
    Harold Bloom

Identity and Art in Frederick Douglass's
Narrative / 7
    Albert E. Stone

Animal Farm Unbound / 29
    H. Bruce Franklin

Narration, Authentication,
and Authorial Control / 45
    Robert B. Stepto

Binary Oppositions in Chapter One
of the *Narrative* / 59
    Henry Louis Gates, Jr.

The Text Was Meant to Be Preached / 77
    Robert G. O'Meally

Autobiographical Acts and the Voice
of the Southern Slave / 95
    Houston A. Baker, Jr.

The Problematic of Self in Autobiography:
The Example of the Slave Narrative / 113
    Annette Niemtzow

Language in Slavery / 131
    ANN KIBBEY

Comprehending Slavery: Language and
Personal History in the *Narrative* / 153
    JOHN SEKORA

The Performance of the *Narrative* / 165
    WILLIAM L. ANDREWS

Chronology / 183

Contributors / 189

Bibliography / 191

Acknowledgments / 193

Index / 195

Editor's Note

This book brings together a representative selection of the best modern critical interpretations of the *Narrative of the Life of Frederick Douglass, an American Slave, Written by Himself* (1845). The critical essays are reprinted here in the chronological sequence of their original publication. I am grateful to Caroline Rebecca Rogers and Henry Finder for their assistance in editing this volume.

My introduction centers upon the role of Southern erotic sadism in the psychic economy of slaveholding, as depicted by Douglass. Albert E. Stone begins the chronological sequence of criticism with his early tribute to the *Narrative* as a precursor to such major black autobiographies as *Black Boy* and *I Know Why the Caged Bird Sings*. H. Bruce Franklin explores the function of animal imagery in representing the brutalizing effects of slavery on both its victims and perpetrators.

The artistic representation of black identity is the subject of Robert B. Stepto, who outlines the structure of the *Narrative* as one of three phases of slave narrative. Douglass's first chapter, in a reading by Henry Louis Gates, Jr., is explored as an instance of the pattern of binary oppositions, as set forth by Levi-Strauss, with Douglass himself mediating and reversing the structuring polarities of the slavocracy.

Robert G. O'Meally examines the relation of the *Narrative* to the Afro-American sermon, while Houston A. Baker, Jr. essentially sees Douglass transforming slave experience into an "autobiographical act" through an uncritical acceptance of the liberating function of literacy itself. In Annette Niemtzow's analysis, the slave narrative is also seen as a step towards freedom, yet a step only, impeded by the cage of an imposed form.

Ann Kibbey, in an advanced rhetorical exegesis of the *Narrative*, shrewdly concludes that the "linguistic virtuosity of the slave who survived must have been impressive." Arguing against the consensus, John Sekora maintains that the work should be viewed not as autobiography at all but rather as "the first, comprehensive personal history of American slavery." This book ends with William L. Andrews's vision of Douglass as "the country's black Jeremiah," working to bring on the Civil War.

Introduction

A rereading of the *Narrative of the Life of Frederick Douglass* gives the impression that the book could have been called *A Slave Is Being Beaten*. The *Narrative* is clearly Douglass's best writing, though wildly uneven, and has considerable force as a fragment of autobiography. Setting aside moral and historical considerations, the book's chief strength is its frightening insight into the erotic psychology of slaveholding, which it exposes as not being wholly unlike the sexual motivations for running a death camp. The peculiar institution of the American South is revealed as a grand version of the economic problem of sadism, or the vicissitudes of the Southern drive. Douglass's poignant account of his relation to his mother is at once the most eloquent paragraph in the *Narrative* and the deliberate introduction to the theme of white sadism that will unify the work:

> I never saw my mother, to know her as such, more than four or five times in my life; and each of these times was very short in duration, and at night. She was hired by a Mr. Stewart, who lived about twelve miles from my home. She made her journeys to see me in the night, travelling the whole distance on foot, after the performance of her day's work. She was a field hand, and a whipping is the penalty of not being in the field at sunrise, unless a slave has special permission from his or her master to the contrary—a permission which they seldom get, and one that gives to him the proud name of being a kind master. I do not recollect of ever seeing my mother by the light of day. She was with me in the night. She would lie down with me, and get me to sleep, but long before I waked she was gone. Very little communication ever took place be-

> tween us. Death soon ended what little we could have while she lived, and with it her hardships and suffering. She died when I was about seven years old, on one of my master's farms, near Lee's Mill. I was not allowed to be present during her illness, at her death, or burial. She was gone long before I knew any thing about it. Never having enjoyed, to any considerable extent, her soothing presence, her tender and watchful care, I received the tidings of her death with much the same emotions I should have probably felt at the death of a stranger.

"She was with me in the night," but whipped if not in the field at dawn. Douglass's white father, perhaps his master, is even more an absence, except as a universal image of guilt and sorrow, master and father coming together as a composite image of the American Inferno:

> Called thus suddenly away, she left me without the slightest intimation of who my father was. The whisper that my master was my father, may or may not be true; and, true or false, it is of but little consequence to my purpose whilst the fact remains, in all its glaring odiousness, that slaveholders have ordained, and by law established, that the children of slave women shall in all cases follow the condition of their mothers; and this is done too obviously to administer to their own lusts, and make a gratification of their wicked desires profitable as well as pleasurable; for by this cunning arrangement, the slaveholder, in cases not a few, sustains to his slaves the double relation of master and father.
>
> I know of such cases; and it is worthy of remark that such slaves invariably suffer greater hardships, and have more to contend with, than others. They are, in the first place, a constant offence to their mistress. She is ever disposed to find fault with them; they can seldom do any thing to please her; she is never better pleased than when she sees them under the lash, especially when she suspects her husband of showing to his mulatto children favors which he withholds from his black slaves. The master is frequently compelled to sell this class of his slaves, out of deference to the feelings of his white wife; and, cruel as the

deed may strike any one to be, for a man to sell his ow
children to human flesh-mongers, it is often the dictate of
humanity for him to do so; for, unless he does this, he
must not only whip them himself, but must stand by and
see one white son tie up his brother, of but few shades
darker complexion than himself, and ply the gory lash to
his naked back; and if he lisp one word of disapproval, it
is set down to his parental partiality, and only makes a bad
matter worse, both for himself and the slave whom he
would protect and defend.

Douglass's rhetoric, though a touch uncontrolled, has just enough irony to qualify as authentic literary language, particularly whenever the issue is Southern Protestantism, then a most peculiar variety of Christianity. The memorable personages in the *Narrative* are those great Protestants of the lash: Plummer, Severe, Gore, Lanman, Bondly, Auld, and Covey, a sevenfold whose very names are suggestive of Dickensian villains. These seven reverberate in the reader's recollection as being worthy of the defendants' dock at Nuremberg or Jerusalem. The *Narrative* is a catalog of the victims of atrocities, from Douglass's Aunt Hester, whose master "would whip her to make her scream, and whip her to make her hush," through Demby, whose "mangled body sank out of sight, and blood and brains marked the water where he had stood," on to Douglass's wife's cousin, a girl of fifteen, beaten to death by Mrs. Hicks for falling asleep while baby-sitting. As this litany of horrors goes on, the almost numbed reader comes to see that Douglass's tone is remarkably temperate for someone whose childhood memories begin with the trauma of his aunt's shrieks of pain: "It was the blood-stained gate, the entrance to the hell of slavery, through which I was about to pass."

Skepticism arises when readers wonder at the economic losses sustained by slaveholders by the mutilation and even extinction of so much valuable property. Douglass, constantly and grimly aware of the slaveholders' balance between the erotic pleasures of sadism, and the commercial displeasures consequent upon acute gratification, is shrewdly persuasive at depicting how business virtue yields to the perversions of the psyche. His psychological insight is most frightening in his account of the decline of Sophie Auld, initially "a woman of the kindest heart and finest feelings." Upon marrying into slave-owning, Mrs. Auld soon enough becomes a demon, poisoned by "irresponsible power." She joins those who, in the parodistic verses that end the *Narrative,* "lay up treasures in the sky, /By making switch and cowskin fly."

Freud held very complex and subtle theories as to the origins of sadism, and his final speculations upon the relation of sadism to the death drive, beyond the pleasure principle, seem to me the least convincing of his speculations in this dark area. But the *Narrative* is very late Freudian in its vision of the deathliness of the masters. Still, the earlier Freudian explanations of sadism are also served by Douglass's unflinching rendition of the terrors that he himself had witnessed. Severe, Gore, Covey and the other real versions of Simon Legree lash away in order to introject the image of the father, earthly and heavenly, source of all authority and righteous dealing:

> I find, since reading over the foregoing Narrative that I have, in several instances, spoken in such a tone and manner, respecting religion, as may possibly lead those unacquainted with my religious views to suppose me an opponent of all religion. To remove the liability of such misapprehension, I deem it proper to append the following brief explanation. What I have said respecting and against religion, I mean strictly to apply to the *slaveholding religion* of this land, and with no possible reference to Christianity proper; for, between the Christianity of this land, and the Christianity of Christ, I recognize the widest possible difference—so wide, that to receive the one as good, pure, and holy, is of necessity to reject the other as bad, corrupt, and wicked. To be the friend of the one, is of necessity to be the enemy of the other. I love the pure, peaceable, and impartial Christianity of Christ: I therefore hate the corrupt, slaveholding, women-whipping, cradle-plundering, partial and hypocritical Christianity of this land. Indeed, I can see no reason, but the most deceitful one, for calling the religion of this land Christianity. I look upon it as the climax of all misnomers, the boldest of all frauds, and the grossest of all libels. Never was there a clearer case of "stealing the livery of the court of heaven to serve the devil in." I am filled with unutterable loathing when I contemplate the religious pomp and show, together with the horrible inconsistencies, which every

where surround me. We have men-stealers for ministers, women-whippers for missionaries, and cradle-plunderers for church members. The man who wields the blood-clotted cowskin during the week fills the pulpit on Sunday, and claims to be a minister of the meek and lowly Jesus. The man who robs me of my earnings at the end of each week meets me as a class-leader on Sunday morning, to show me the way of life, and the path of salvation. He who sells my sister, for purposes of prostitution, stands forth as the pious advocate of purity. He who proclaims it a religious duty to read the Bible denies me the right of learning to read the name of the God who made me. He who is the religious advocate of marriage robs whole millions of its sacred influence, and leaves them to the ravages of wholesale pollution. The warm defender of the sacredness of the family relation is the same that scatters whole families,—sundering husbands and wives, parents and children, sisters and brothers,—leaving the hut vacant, and the hearth desolate. We see the thief preaching against theft, and the adulterer against adultery. We have men sold to build churches, women sold to support the gospel, and babies sold to purchase Bibles for the *poor heathen! all for the glory of God and the good of souls!* The slave auctioneer's bell and the church-going bell chime in with each other, and the bitter cries of the heart-broken slave are drowned in the religious shouts of his pious master. Revivals of religion and revivals in the slave-trade go hand in hand together. The slave prison and the church stand near each other. The clanking of fetters and the rattling of chains in the prison, and the pious psalm and solemn prayer in the church, may be heard at the same time. The dealers in the bodies and souls of men erect their stand in the presence of the pulpit, and they mutually help each other. The dealer gives his blood-stained gold to support the pulpit, and the pulpit, in return, covers his infernal business with the garb of Christianity. Here we have religion and robbery the allies of each other—devils dressed in angels' robes, and hell presenting the semblance of paradise.

We have our contemporary instances of Douglass's quiet and strong sentence: "Revivals of religion and revivals in the slave-trade go hand in hand together." Our current Frederick Douglass is James Baldwin, and his essays provide adumbrations of Douglass's emphasis. The psychic introjection of the image of the father requires a blood-sacrifice, and such ritual still goes on among us. Melanie Klein grimly observes that: "The early stages of the Oedipus conflict are dominated by sadism." Her general conclusion as to our universal "sadistic appropriation and exploration of the mother's body and of the outside world" is amply supported by Douglass's epic *Narrative*, with its procession of suffering substituting for the Southern mother's body.

Identity and Art in Frederick Douglass's *Narrative*

Albert E. Stone

"America has the mournful honor of adding a new department to the literature of civilization—the autobiographies of escaped slaves." This announcement by the Reverend Ephraim Peabody, a New Bradford minister and abolitionist, appeared in the *Christian Examiner and Religious Miscellany* for July 1849, prefacing a long discussion of five slave narratives which had been published during the preceding four years. The personal histories were those of Henry Watson, Lewis and Milton Clarke, William Wells Brown, Josiah Henson, and Frederick Douglass. "We place these volumes without hesitation among the most remarkable productions of the age—" Peabody continued, "remarkable as being pictures of slavery by the slave, remarkable as disclosing under a new light the mixed elements of American civilization, and not less remarkable as a vivid exhibition of the force and working of the native love of freedom in the individual mind." This appreciation of the emotional power and cultural significance of slave narratives was indeed prophetic. Though successors have widened his frame of reference and modified some of his genteel judgments, Ephraim Peabody remains one of the first white critics to pay serious attention to a new form of autobiography in America. Some years later in 1863, one of the new black writers he had discussed became himself an annalist of the Negro. In *The Black Man: His Antecedents, His Genius, and His Achievements* William

---

From *CLA Journal* 17, no. 2 (December 1973). ©1973 by the College Language Association.

Wells Brown cited these narratives as the first black voices in American literature. Moreover, like Peabody, Brown singled out Frederick Douglass as the master of this new literature. "The narrative of his life, published in 1845, gave a new impetus to the black man's literature," he wrote. "All other stories of fugitive slaves faded away before the beautifully written, highly descriptive, and thrilling memoir of Frederick Douglass." Peabody and Brown announce early what history has since confirmed: the *Narrative of Frederick Douglass, an American Slave, Written by Himself* is at once an important cultural document and an unusual work of autobiographical art.

By 1849 the slave narrative had already become one of the more popular forms of political literature in the North. Peabody reinforced but did not create Douglass's fame. The *Narrative* had already gone through seven editions and Benjamin Quarles has estimated that by 1850 it had sold some 30,000 copies here and in the British Isles. Brown's narrative had sold 8,000 copies by 1849 and Henson's *Life* was soon to become even more famous—and notorious—as a result of the publicity linking him as "the original Uncle Tom" to Harriet Beecher Stowe's best-seller; within Henson's lifetime the three versions of his autobiography would sell 100,000 copies. Well before the appearance of *Uncle Tom's Cabin,* thousands of American, Canadian, and British readers had already formed impressions of chattel slavery in the Southern states by reading these personal histories as they appeared in magazines, in twenty-five-cent pamphlets, and in books costing a dollar or a dollar and a half. As Charles H. Nichols points out in *Many Thousand Gone,* the definitive history of the slave narrative, these were the first American autobiographies widely read by a popular audience—and for some of the same reasons which have made *The Autobiography of Malcolm X* a best-seller today.

Since the slave narrative flourished in close connection to the abolition movement and appeared (and declined) chiefly in the three decades before the Civil War, the modern reader tends to be concerned, as Nichols is, with the historical context of these books—their composition and publication, their reception and impact, their claims to historical truth or accuracy. Douglass and his fellow fugitives did indeed create an important literature of protest and propaganda. But to assert this is also to recognize that historicity cannot be divorced from other considerations equally important in assessing the permanent cultural value of these works—consideration of literary style and rhetorical strategy, of psychological revelation and motivation. As autobiographies, the *Narrative* and other similar works occupy the territory between history and art, biography and fiction, memory and imagination. When the ex-slave asked the question (or was urged to do so by a white sponsor or collaborator) which all autobiographers ask: "Why am I writing the story of my life?" the immediate answer was plain: to describe the experience of being a chattel and then *not* being one so vividly that the white reader would be moved to destroy the oppressive institution. To this end, the most effective means was to create a convincing impression of historical veracity and verisimilitude. Thus the editor of Douglass's second autobiography, *My Bondage and My Freedom,* declared: "the reader's attention is not invited to a work of art, but to a work of FACTS. There is not a fictitious name or place in the whole volume . . .; every transaction therein described actually transpired." Telling the unvarnished truth about verifiable experience and re-creating thereby the self in relation to time, history and change, is an aim of all authentic memoirs, but one which had a particular value for the writer and editor of slave narratives.

Yet all history is, as J. H. Hexter has shown, a deliberate artistic creation. Slave narratives like Douglass's exhibit a variety of literary devices for recording a past, persuading belief, and motivating action. Capitalizing FACTS above is one simple instance of such a rhetorical tactic. Other techniques were devised for the strategy of "sticking to the facts," for in recording the bare details of life as a slave—including the pathos and tragedy of slave auctions and family separations, and drama and excitement of escape to freedom—the writer could hardly avoid the appearance of fiction or the atmosphere of melodrama. Thus the line between autobiography and fiction became a fine one, as is suggested not only by the title of Josiah Henson's second autobiography, *Truth Stranger Than Fiction* (1858), but also by the early appearance of actual romances or pseudohistories like Richard Hildreth's *The Slave: or Memoirs of Archy Moore* (1836). The later novels of Mrs. Stowe and William Wells Brown derived much of their force from the reader's realization that actual life histories existed to authenticate what the novelist had imagined.

If historical truth could have an effect stronger and stranger than fiction, one way to achieve this effect was not to explore the whole system of slavery but instead to exploit the natural focus of autobiography upon private experience and the single self. This, too, provided a fiction-like perspective. How the individual slave became a man in the act of escape was both plot and moral of the slave narrative. In this respect Douglass's *Narrative* is the exemplary work in the genre. By forging a portrait of himself, rather than simply writing history or abolitionist propaganda, Douglass reveals himself a true autobiographer. He also distinguishes his achievement from that of other ex-slave writers like Harriet Jacobs, author of *Incidents in the Life of a Slave Girl* (1861) and Charles Ball, author of *Fifty Years in Chains* (1837). Harriet Jacobs, with assistance from Lydia Maria Child, dramatizes the experience of slavery by means of fictional names, dialogue, sentimental language, and a melodramatic plot of fear, seduction, and flight; hers is personal history under the influence of sentimental romance. Ball's account, on the other hand, reads more like history than autobiography, for its tone and perspective draw attention away from the narrator and his developing identity, toward the generalized facts of life under the brutalizing institution. More successfully than either of these, Douglass saw and exploited the crucial difference between autobiography and its allied forms, history and fiction. As I shall seek to demonstrate, he would have agreed with modern critics of autobiography like F. R. Hart who emphasize the distinctive aim of autobiography. Hart observes that "in understanding fiction one seeks an imaginative grasp of another's meaning; in understanding personal history one seeks an imaginative comprehension of another's historic identity. 'Meaning' and 'identity' are not the same kind of reality and do not make the same demands." Identity through history and art, self as the container of meaning—in these terms, I would argue, lies a proper understanding of Frederick Douglass's *Narrative*.

Though later readers have followed Peabody and Brown in accepting the preeminence of this slave narrative, nevertheless the true artistry of the *Narrative* has yet to be fully analyzed and appreciated. This is surprising in light of Douglass's fame as a public figure, which has been recorded in several biographies. Among the critics and literary scholars who have contributed to a richer realization of Douglass's achievements as writer are Vernon Loggins, Benjamin Brawley, Arna Bontemps, Charles H. Nichols, Benjamin Quarles, Jean F. Yellin, and Houston A. Baker, Jr. Each has illuminated certain aspects of the *Narrative;* none, however, has exhausted its deceptive richness of language, style, and structure. A typical recent discussion is Jean Fagan Yellin's in *The Intricate Knot*. It is, she affirms, "a classic

American autobiography" with a narrative style which combines sparcity and aptness of symbolic detail with a dramatic pace and structure. "Douglass' *Narrative* is not a flawless work of art," she concludes, "but it expresses more than the boundless incident and passion of the other slave autobiographies and of the contemporary plantation and abolitionist fiction." Unexceptionable as these judgments are, they occur in a three-page commentary—too brief to do justice to their implications. A fuller discussion is Quarles's introduction to the John Harvard Library edition of the *Narrative,* but this essay never gets around to the closer look at narrative strategy and style which is promised; the historical and biographical background takes up most of the space. The fullest, most sensitive reading which has so far appeared is Houston Baker's in *Long Black Song.* Even this shrewd analysis fails, however, to define precisely and explore adequately all its insights. Baker is correct in seeing the *Narrative* as "sophisticated literary autobiography" and he admirably describes the characteristics of Douglass's prose style—the understated, visualized narrative, the dry, humane irony, the deft characterizations, the adroit use of animal imagery, antithesis, and the agrarian setting. But Baker's linkage of Douglass as a "spiritual" autobiographer to the tradition of Mather, Franklin, and Henry Adams is debatable on several points, for he admits Douglass is never centrally preoccupied with inner experiences of conversion, salvation, or confession. As the appendix indicates, Douglass's Christianity was a practical, public, moral matter. Baker also sees Douglass as "something of a mythic figure" but fails to suggest evidence in the text—rather than in the minds of black or white readers—for this self-mythification; the weight of Baker's discussion, in fact, works against a mythic reading of the *Narrative.* Moreover, Baker simplifies the movement "from a cruel physical bondage to freedom" narrated and psychologically explored in Douglass's book. The achievement of a prior, inner freedom in the fight with Edward Covey is recognized but unrelated to other episodes and patterns of metaphor. Nevertheless, Baker's provocative discussion opens new issues and suggests the need for a careful examination of Frederick Douglass as narrative artist and artificer of the self. To look at the *Narrative* in these terms—at once literary, historical, and psychological—should not detract from the author's polemical purposes. For the more clearly and fully we see the man *and* the writer—the man revealed in the act of discovering and recreating his own identity—the more we acknowledge the force of his argument for an end to slavery's denial of individuality and creativity.

I

Douglass's identification of self begins with the title and prefaces, which establish conditions of the autobiographical contract between black writer and white audience. *Narrative of the Life of Frederick Douglass, an American Slave, Written by Himself* has a directness later replaced by the more figurative title *My Bondage and My Freedom,* in the 1855 version and then in 1892 by the final, historical *Life and Times of Frederick Douglass.* Probably the most meaningful part of the title to the reader of 1845 was *Written by Himself.* The phrase reverberates with *An American Slave* to suggest the poles of Douglass's experience—his past a dependent slave, his present as independent author. The practical need for these phrases was doubtless the attacks in the pro-slavery press on the authenticity of slave narratives as not simply biased but untrue because ghostwritten by white abolitionists who knew nothing of slavery. Like the earlier case of the *Narrative of James Williams,* Douglass's *Narrative* was labeled a fraud soon after publication. In a letter to the Delaware *Republican* A. C. C. Thompson challenged the author and publishers. "About eight years ago, I knew this recreant slave by the name of Frederick Bailey (instead of Douglass)," wrote the ex-slaveholder.

> He then lived with Mr. Edward Covey, and was an unlearned, rather an ordinary negro, and am confident he was not capable of writing the Narrative alluded to; for none but an educated man, and one who had some knowledge of the rules of grammar, could write so correctly; although to make the imposition at all creditable, the composer has labored to write it in as plain a style as possible; consequently the detection of this first falsehood proves the whole production to be most notoriously untrue.

Douglass's rejoinder, in the 1846 English edition of the *Narrative,* made effective use of Thompson's charge to stress the central theme of his story—self-transformation. "You are confident I did not write the book," he observed,

> the reason of your confidence on this point is, that I was, when you knew me, an unlearned and rather ordinary ne-

gro. Well, I have to inform you, that you knew me under very unfavorable circumstances;. . . For if any one had told me seven years ago that I should ever be able to *dictate* such a Narrative, to say nothing of *writing* it, I should have disbelieved the prophesy. I was then a mere wreck; Covey had beaten and bruised me so much, that my spirit was crushed and broken. Frederick the Freeman is very different from Frederick the Slave. . . . Freedom has given me a new life.

Whereas Douglass's title asserts the identity and responsibility of its black author, the first pages of the *Narrative* are devoted to guarantees by white sponsors. The preface and introduction by William Lloyd Garrison and Wendell Phillips are double assurances by two of abolitionism's greatest names of the book's authenticity. Virtually every nineteenth-century slave narrative carried such seals of white approval. Indeed, the practice has persisted long after Emancipation, as Dorothy Canfield Fisher's introduction to *Black Boy* (1945) and M. S. Handler's to *The Autobiography of Malcolm X* (1965) both attest. At the time Douglass wrote the practice was well established; one principal purpose was to state openly the circumstances of authorship and the degree of editorial assistance. Despite accusations from anitabolitionists, these accounts, with their sometimes condescending but explicit introductions, are less dishonest forms of American autobiography than many present-day ghostwritten lives of Hollywood or SuperBowl celebrities.

What immediately distinguishes the *Narrative* from most other slave accounts is Douglass's skill in using the introductions for his own purposes, so that what is elsewhere an extraneous essay becomes part of a unified form. The first advantage, of course, is a dramatic presentation of himself by another, thus dealing at once with the reader's possible imputation of vanity. "In the month of August, 1841," Garrison begins,

> I attended an anti-slavery convention in Nantucket, at which it was my happiness to become acquainted with FREDERICK DOUGLASS, the writer of the following Narrative. . . . I shall never forget his first speech at the convention—the extraordinary emotion it excited in my own mind—the powerful impression it created upon a crowded auditory, completely taken by surprise—the ap-

plause which followed from the beginning to the end of his felicitous remarks.

As Garrison proceeds to praise Douglass also as a writer his own style as writer-orator becomes sharply contrasted to that of his black protegé's. "Mr. Douglass has very properly chosen to write his own Narrative, in his own style," he continues,

> it is therefore, entirely his own production; . . . He who can peruse it without a tearful eye, a heaving breast, an afflicted spirit,—without being filled with an unutterable abhorrence of slavery and all its abettors, and animated with a determination to seek the immediate overthrow of that execrable system,—without trembling for the fate of this country in the hands of a righteous God, who is ever on the side of the oppressed, and whose arm is not shortened that it cannot save,—must have a flinty heart, and be qualified to act the part of a trafficker "in slaves and the souls of men."

An introduction has turned into a speech and the personal subject largely lost in oratorical emotion. Though briefer, Wendell Phillips, too, falls into similar pulpit language and righteous regional feeling. These fulsome outpourings are deftly counterpointed by Douglass's own style and language throughout the *Narrative*. His final paragraph brings the reader back full circle to Garrison's opening one, but the Nantucket event is now re-created with the quiet authority of his own and not the white man's voice:

> I had not long been a reader of the "Liberator," before I got a pretty correct idea of the principles, measures and spirit of the anti-slavery reform. I took right hold of the cause. . . . I seldom had much to say at the meetings, because what I wanted to say was said so much better by others. But, while attending an anti-slavery convention at Nantucket, on the 11th of August, 1841, I felt strongly moved to speak, and was at the same time much urged to do so by Mr. William C. Coffin, a gentleman who had heard me speak in the colored people's meeting at New Bedford. It was a severe cross, and I took it up reluctantly. The truth was, I felt myself a slave, and the idea of speaking to white people weighed me down. I spoke but a few moments, when I felt a degree of freedom, and said what

Identity and Art in Frederick Douglass's Narr[ative]

> I desired with considerable ease. From that time until no[w] I have been engaged in pleading the cause of my bre[th]ren—with what success, and with what devotion, I leave those acquainted with my labors to decide.

Here the whole movement of the autobiography is succintly recapitulated—the desire for freedom but the sense of being a slave, speaking out as the symbolic act of self-definition, Douglass's quiet pride in his public identity. Though the appendix apparently undercuts the symmetry of this ending, this afterthought on religious hypocrisy also asserts his independence of official white institutions. Thus the original contrast is maintained throughout: while the white men Garrison and Phillips, argue a cause and point to this extraordinary black man as proof, Douglass's own account creates the image of a man, and this act of identity authenticates the cause of abolition.

The process from first to last is the creation of an *historical* self. "I was born in Tuckahoe, near Hillsborough, and about twelve miles from Easton, in Talbot County, Maryland." So begins his story, which ends on an equally matter-of-fact note: "I subscribe myself, FREDERICK DOUGLASS. Lynn, Mass., April 28, 1845." Both statements sound flatly conventional but carry a special meaning. Under slavery, man possesses no such historic identity as name, date, place of birth or residence usually provide. Douglass has achieved these hallmarks of historicity, has attached himself to time, place, society. Therefore he shows no wish to escape from history. As soon as memory provides them, and it is safe to do so, he gives names, dates, titles, places—all the usual evidence of existence which many slaves are denied. Yet Douglass never loses himself in memoir, as do many slave narrators, by making his account merely factual or typical. To be sure, the *Narrative* records many experiences and emotions shared by other fugitive slaves, but these are stamped with Douglass's own imagination.

This individual vision develops gradually, but can be seen even in the first primal scene, the flogging of Aunt Hester, "a woman of noble form, and of graceful proportions" who has aroused the passions and ire of the master. Douglass terms his initiation "the blood-stained gate, the entrance to the hell of slavery, through which I was about to pass,"—apt imagery for the violent emotions of master, slave woman, and the terrified child in the closet. The metaphor of "blood-stained gate" is typical of Douglass's language. Traditionally

Christian on one level, it also communicates more private and inchoate feelings about birth, sexuality, violence, dark mothers and white fathers. To deal with such emotions and forces the boy has little of the family love or religious consolation available to other ex-slave writers like J. W. C. Pennington or Solomon Northup. Instead, like many slave children, he can recall nothing of his father except that he was reportedly white, and remembers seeing his mother only by night, for she lived on another plantation. The child's sense of isolation and his ultimate response are both neatly connected in his explanation: "It is a common custom," he observes dryly, "in the part of Maryland from which I ran away, to part children from their mothers at a very early age." His later experiences—as house servant and field hand, in Baltimore and on plantations large and small, with brutal masters and some kind ones—continue his personal history in terms also representative. However, Douglass prefers the personal and seldom goes out of his way to dramatize situations which his readers are expecting but which are not actually part of his own remembered past. When he does refer to the sufferings of other slaves, these are carefully identified. The result is a narrative with less violence but more authority than many works in this genre.

The gradual enlargement of perspective in the *Narrative* is made natural and appropriate by Douglass's autobiographical point of view. He does not limit himself to the growing child's impressions but, like Benjamin Franklin (with whom, Alain Locke has noted, he has several parallels), writes both as experiencing boy and experienced adult. This double vision is managed with considerable skill throughout the book. On the opening page he contrasts himself as a young slave whose only birthday is an animal's—"planting-time, harvest-time, cherry-time, spring-time, or fall-time"—with the grown writer whose present identity shares the anonymity and ignorance of slavery. "The nearest estimate I can give makes me now between twenty-seven and twenty-eight years of age. I come to this, from hearing my master say, some time during 1835, I was about seventeen years old." Like Malcolm X and Claude Brown, Douglass is a very youthful autobiographer, with a young man's vivid memories instead of an older writer's diaries or reminiscences as resources.

Other means of juxtaposing past and present selves in order to dramatize change and continuity are even more arresting and effective. Speaking of his childish sufferings, he remarks:

> I had no bed. I must have perished with cold, but that, the coldest nights, I used to steal a bag which was used for carrying corn to the mill. I would crawl into this bag, and there sleep on the cold, damp, clay floor, with my head in and feet out. My feet have been so cracked with the frost, that the pen with which I am writing might be laid in the gashes.

Still more striking is the memorable description of the slaves' songs he remembers hearing on the road to the Great House Farm. He writes first from his present perspective. "I have often been utterly astonished, since I came to the north, to find persons who could speak of the singing, among slaves, as evidence of their contentment and happiness," he remarks. Then he recaptures his past emotion: "those wild notes always depressed my spirit, and filled me with ineffable sadness." Finally he returns to the present to drive home his point: "I did not, when a slave, understand the deep meaning of those rude and apparently incoherent songs. I was myself within the circle; so that I neither saw nor heard as those without might see and hear." His message is clear, but more complex than with most slave narrators. Neither the slaves themselves nor a sympathetic outsider—like, say, the sympathetic English actress Fanny Kemble who could not fathom the significance of the slave singing on her Georgia plantation—is in a position to tell the truth about this music. Only by being *black* and *becoming* free has Douglass earned the rank and right of interpreter. It is a message whose precision Stephen Crane would have understood.

Douglass's departure from Colonel Lloyd's plantation provides another occasion for dramatizing the double perspective. Embarking for Baltimore, the young boy, who symbolically scrubbed all the dead skin from his knees and donned his first pair of trousers, placed himself "in the bows of the sloop, and there spent the remainder of the day in looking ahead, interesting myself in what was in the distance rather than in things near by or behind." Only in retrospect does the traveler see the significance of this preliminary escape to the city, which opens another "gateway, to all my subsequent prosperity." Contrasting that boy "in the galling chains of slavery" with himself now "seated by my own table . . . writing this Narrative," he nevertheless affirms, as do all true autobiographers, a deep continuity between the two selves. "From my earliest recollection," he writes, "I date the entertainment of a deep conviction that slavery would not always be able to hold me within its foul embrace." This same sense of himself as two persons yet one self is likewise expressed in Douglass's various names and aliases; though at different times he becomes Bailey, Johnson, and Douglass, he never relinquishes Frederick. "I must hold on to that, to preserve a sense of my identity," he declares near the close of the *Narrative*.

Among later episodes which express this writer's evolving identity, the crucial ones are his learning to read and write. Expression is at the core of selfhood for Frederick Douglass. In Baltimore, Mrs. Auld's assistance is soon halted by her husband, but young Frederick is not daunted. Their white repression awakened "sentiments within that lay slumbering," and he turned to the white boys of the street for help in reading. "When I was sent of errands," he relates in language that suggests Franklin," I always took my book with me, and by going one part of my errand quickly, I found time to get a lesson before my return. I used to also carry bread with me. . . . This bread I used to bestow upon the hungry little urchins who, in return, would give me that valuable bread of knowledge." The climax of this process of discovery came when he was twelve, at a time when "the thought of being *a slave for life* began to bear heavily upon my heart." Then he discovered the *Columbian Orator*. This book becomes a key link between boy and man, slave and abolitionist, for its full *The Columbian Orator: Containing a Variety of Original and Selected Pieces; Together with Rules; Calculated to Improve Youth and Others in the Ornamental and Useful Art of Eloquence* by Caleb Bingham. "Among much of other interesting matter," Douglass recalls, "I found in it a dialogue between a master and his slave." He continues: "In the same book, I met with one of Sheridan's mighty speeches on and in behalf of Catholic emancipation. These were choice documents to me. I read them over and over again with unabated interest." Though memory has played him slightly false (the actual speech is not by Sheridan but is a *Speech in Irish Parliament by O'Connor in Favor of Roman Catholic Emancipation, 1795*) a glance at its contents proves the wisdom of masters and mistresses in trying to keep such books from the eyes of slaves. Here, for instance, the boy read a *Discourse on Manumission of Slaves* by the Rev. Samuel Miller, *Slaves in Barbary: A Drama in Two Acts* by Everett, and a dialogue about civilization between an Indian and a white man. The most suggestive excerpt, perhaps, to the young reader was the one remembered, the *Dialogue between A Master and a Slave* by Aiken, which appears just before O'Connor's speech. Here is a sample of the ideas the boy encountered:

> MASTER. Now, villian [sic]: What have you to say for this second attempt to run away? . . .
>
> SLAVE. I am a slave. That is answer enough.
>
> MASTER. I am not content with that answer. I thought I discerned in you some tokens of a mind superior to your condition. I treated you accordingly. You have been comfortably fed and lodged, not over-worked, and attended with the most humane care when you were sick. And is this the return? . . .
>
> SLAVE. Providence gives [the robber] a power over your life and property. . . . But it has also given me the legs to escape with; and what should prevent me from using them? . . . Look at these limbs, are they not those of a man? Think that I have the spirit of a man too.

After the Master has freed his Slave, the latter addresses him:

> Now I am indeed your servant, though not your slave. And as the first return I can make for your kindness, I will tell you freely the condition in which you live. You are surrounded with implacable foes, who long for a safe opportunity to revenge upon you and the other planters all the miseries they have endured. . . . You can rely on no kindness on your part, to soften the obduracy of their resentment. Your have reduced them to the state of brute beasts; and if they have not the stupidity of beasts of burden, they must have the ferocity of beasts of prey. Superiour force alone can give you security. . . . Such is the social bond between master and slave!

In the *Narrative* itself one may see the ultimate effect on the young boy of discovering how the aspirations and realities of his slave's life could find adequate expression. "The reading of these documents enabled me to utter my thoughts," he recalls with characteristic understatement.

But to reach this point he needed to write. This decisive step towards his present identity as a free black man occurred under circumstances which are recollected in detail:

> The idea as to how I might learn to write was suggested to me by being in Durgin and Bailey's ship-yard, and frequently seeing the ship carpenters, after hewing, and getting a piece of timber ready for use, write on the timber the name of that part of the ship for which it was intended.

Specific details like these have a deceptive simplicity. Learning to write in a shipyard bearing in part his own name is both an historical and a symbolic event. In recording an actual occurrence, one which connects the twelve-year-old boy to the present writer in Lynn, he continues a pattern of event and image linked together to articulate his autobiographical identity. For Douglass's association of learning to read and write with ships and shipyards is not accidental. It recalls earlier and later moments when boy and man are seen in terms of ships, shipbuilding, and sailing across the water. We have already noted the first such occasion—placing himself in the very bow of the sloop sailing towards Baltimore and eventual freedom. Another moment is the famous apostrophe to the ships on the Chesapeake:

> Those beautiful vessels, robed in purest white, so delightful to the eye of free-men, were to me so many shrouded ghosts, to terrify and torment me with thoughts of my wretched condition. . . . I would pour out my soul's complaint, in my rude way, with an apostrophe to the moving multitude of ships:—
> You are loosed from your moorings, and are free; I am fast in my chains, and am a slave! . . . You are freedom's swift-winged angels, that fly round the world; I am confined in bands of iron!

Beneath this awkward rhetoric are some powerful personal associations linking ships and sails not simply to freedom, adventure, and literacy, but also the color white and the word "angel" with Mrs. Hugh Auld, his white preceptress who started him on the voyage to a free self and then betrayed him. In less emotional language, Douglass records another event in this complex when he describes his last act of self-assertion as a Maryland slave—his fight in Gardner's shipyard with the white apprentices. In his memory and imagination Douglass identifies freedom with both learning from and fighting with whites; both relationships are often associated with ships or their construction. Thus though we do not learn so in the *Narrative,* but only later in *Life and Times,* it is fitting that this young plantation slave escaped to the North disguised as a sailor. What the *Narrative* does tell us is that an Irishman on a wharf asked him first the vital question: "Are ye a slave for life?" Here manifestly is a rich mixture of persons, places, sights, acts, and emotions which have combined in the autobiographer's memory to become what James Olney would call a "metaphor of self." Douglass's deepest impulses towards freedom, personal identity, and self-expression are fused and represented in these memories and images of ships and the sea. Therefore it is wholly appropriate that the final act by which selfhood is confirmed in the *Narrative* is speaking at the meeting on the island of Nantucket. Far more so than animal imagery, I believe, this pattern is central to Frederick Douglass's first autobiography for it connects and defines all stages of his personal history. "The following of such thematic designs through one's life," writes Vladimir Nabokov in *Speak, Memory,* "should be, I think, the true purpose of autobiography."

II

These literary strategies of self-presentation—the symmetry of Garrison's opening and his own closing paragraphs, the unity provided by the double perspective and by repeated experiences and images of ships, shipbuilding, and sailing across the water—set Douglass's *Narrative* apart from other artful accounts by ex-slaves like Henry Bibb, Solomon Northup, and Harriet Jacobs. But one must not forget that autobiography depends upon memory as much as on imagination. All remembered events do not fit readily into neat structural or imagistic patterns, no matter how many emotional needs are satisfied by trying to make them do so. Furthermore, one should not lose sight of Douglass's polemical purposes or the expectations of his readers. These readers, some of them unsophisticated and many suspicious of too much artistry, would be won over more immediately by a *story* than by a *point of view* or a *pattern of imagery.*

Hence Douglass's emphasis upon exciting narrative, hence his climax in the gripping fight with Edward Covey. This event, he tells us, was the turning-point of his life. It occupies the same central place in the *Narrative*.

"You have seen how a man was made a slave; you shall see how a slave was made a man." Everything in the re-created life of Frederick Douglass builds to and leads away from this declaration. The fight between the sixteen-year old boy and the white farmer occurs in chapter 10—nearly at the end of the *Narrative*. In content, this chapter is a microcosm of the whole *Narrative*. The events described cover exactly a year—1833 in the young man's memory but amended to 1834 in later editions—and thus possess some of the symbolic unity of *Walden*. In becoming the clumsy field hand sent out with the equally clumsy oxen, Douglass is thrust back into the animal's place, easily brutalized there by Covey's whip. But like the oxen he, too, kicks over the lines, will not finally be broken to the yoke. The inspiration to rebel, interestingly enough, is not clearly understood—"from whence came the spirit I don't know" he confesses; "I resolve to fight."—but if follows the sight of the sails on the Chesapeake and derives obvious support from the offer by the superstitious slave Sandy Jenkins of the magical root as a protection. After the fight, which lasts two hours, his transformation is sudden and complete:

> It rekindled the few embers of freedom, and revived within me a sense of my own manhood. . . . It was a glorious resurrection, from the tomb of slavery, to the heaven of freedom. My long-crushed spirit rose, cowardice departed, bold defiance took its place; and I now resolve that, however long I might remain a slave in form, the day had passed forever when I could be a slave in fact.

The remainder of the chapter completes in narrative terms the rebirth here announced. From Covey's hell he moves to the comparative heaven of Mr. Freeland's. The new master's name—like the earlier one, Mr. Severe—is emblematic of Douglass's fortunes. However, a slave's life even under a "good" master cannot be heavenly, as the brutal disruption of his Sabbath school and the betrayal by a fellow slave of his attempted escape both prove. Nonetheless, he affirms, "my tendency was upward." The chapter closes with the young slave back in Baltimore and earning a good wage, which his

Identity and Art in Frederick Douglass's *Narrative* / 23

master appropriates. "The right of the grim-visaged pirate upon the high seas is exactly the same," observes Douglass, and again we note how readily his indignation employs the imagery of the ocean.

As in narrative form and content, so in style is chapter 10 representative of the whole work. It exhibits his two voices with characteristic clarity. The dominant one is the unassuming prose narrator who can set a scene, describe an action, or portray a person with forceful economy. This, for instance, is Edward Covey:

> Mr. Covey's *forte* consisted in his power to deceive. His life was devoted to planning and perpetrating the grossest deceptions. Every thing he possessed in the shape of learning or religion, he made conform to his disposition to deceive. He seemed to think himself equal to deceiving the Almighty. He would make a short prayer in the morning, and a long prayer at night; and, strange as it may seem, few men would at times appear more devotional than he. The exercise of his family devotions were always commenced with singing; and, as he was a very poor singer himself, the duty of raising the hymn generally came upon me. He would read his hymn, and nod at me to commence. I would at times do so; at others, I would not.

Such a description characterizes both the individual and an institution—here the "religion of the south." Thus the balanced antiphonal structure of many of Douglass's sentences is wholly appropriate. A sentence like "The longest days were too short for him, and the shortest nights too long for him" reveals Covey as slave-driver and also as self-driven Southern Protestant, and does so in rhythms strongly reminiscent of the Old Testament, particularly the Psalms. Once attuned to this cadence, the reader recalls how many of the work's aptest aphorisms obey this pattern. "What he most dreaded, that I most desired. What he most loved, that I most hated," describes Hugh Auld's opposition to his learning to read. "I was ignorant of his temper and disposition; he was equally so of mine," is another comment which also reveals the psychological inspiration for this balanced style. The ex-slave sees himself from the start on equal and opposite terms with the white world of slavery. The shape as well as the content of his sentences expresses this equality and energetic opposition. Hence the formal fitness of the *Narrative*'s key sentence: "You have seen how a man was made a slave; you shall see how a slave was made a man." Douglass's story rests and rocks upon that semi-colon.

Douglass has, of course, another voice—the rich periods of the pulpit and platform, which sound so inflated and indulgent to modern ears. However, Alain Locke has warned that Douglass was "by no means the dupe of his own rhetoric." Like his fellow abolitionists, he knew that readers as well as conventioneers expected large doses of sentiment and pathos. In chapter 10 the only instance of this style is the apostrophe to the sailboats. An earlier, lengthier one, which reads almost like a parody of John Pendleton Kennedy's *Swallow Barn,* is the bathetic description (much reduced in later editions) of his grandmother and her solitary cabin. Both passages, despite their fitness for other purposes, sound out of pitch with other parts of the *Narrative* written to more telling emotional effect in Douglass's quieter style. Yet the modern reader must be careful about over-nice judgments of tone and language which miss the emotional depths. Chapter 11 contains, for instance, the last of the purple passages, but this one, because it is backed by the accumulated weight of experience of the whole book, rings truer than earlier outbursts. Here deep and genuine feelings roll irresistibly over the reader. "Let him be a fugitive slave in a strange land—" Douglass exclaims,

> a land given up to be the hunting-ground for slaveholders—whose inhabitants are legalized kidnappers—where he is every moment subjected to the terrible liability of being seized upon by his fellow-men, as the hideous crocodile seizes upon his prey!—I say, let him place himself in my situation—without home or friends—without money or credit—wanting shelter, and no one to give it—wanting bread, and no money to buy it,—and at the same time let him feel that he is pursued by merciless men-hunters, and in total darkness as to what to do, where to go, where to stay,—perfectly helpless both as to the means of defence and means of escape,—in the midst of plenty, yet suffering the terrible gnawings of hunger,—in the midst of houses, yet having no home,—among fellow-men, yet feeling as if in the midst of wild beasts, whose greediness to swallow up the trembling and half-famished fugitive is only equalled by that with which the monsters

of the deep swallow up the helpless fish upon which they subsist,—I say, let him be placed in this most trying situation,—the situation in which I was placed,—then, and not till then, will he fully appreciate the hardships of, and know how to sympathize with, the toil-worn and whip-scarred fugitive slave.

Here, as it seems to me, one experiences what Leo Marx has called the "literary power" of a genuine work of art. The modern reader is prepared to agree with the hopeful editor of the abolitionist *Chronotype* who in 1853 confidently predicted: "This fugitive slave literature is destined to be a powerful lever. We have the most profound conviction of its potency. We see in it the easy and infallible means of abolitionizing the free states. Argument provokes argument, reason is met by sophistry; but the narratives of slaves go right to the hearts of men." Unfortunately, as Charles Nichols pointed out in 1948, the testimony of history does not bear out the editor's belief in the power of these books. Though hearts were indeed moved, minds disabused of much misinformation, and imaginations fired by vivid pictures of slavery and of the black man's actual and potential achievements in coping with slavery, nevertheless American political behavior was not fundamentally altered. Nichols's judgment chastens the enthusiasm of those who believe literary power is readily translated into political action; he points out that only those already predisposed by social, economic, and religious outlook to be open-minded were much affected by personal histories like Douglass's. "One is forced to the conclusion that, though widely read, the narratives effected no vital change in American attitudes," Nichols concludes.

When American hearts were moved, Nichols adds, it was chiefly in a sentimental fashion. Settled convictions about the inferiority of the Negro—beliefs one might expect to be upset by reading so powerful and artful a book as the *Narrative*—were, it appears, seldom changed. If this is true—and questions of the impact of propaganda art on public opinion and behavior are exceedingly difficult to measure—a small but significant factor may be the development towards sentimentality and extreme bathos in the slave narratives published after 1849. The later autobiographies of Douglass himself are, if not representative, as least indicative of a general loss of emotional force and economy. *My Bondage and My Freedom* and the *Life and Times* are not only greatly expanded accounts of a long, distinguished career but are also much looser in style, structure, and imaginative power. Jean Yellin, not the first to note this loss of unity, provides a succinct example by contrasting the key sentence of the 1845 *Narrative* with its 1855 revision: "You have, dear reader, seen me humbled, degraded, broken down, enslaved, and brutalized, and you understand how it was done; now let us see the converse of all this, and how it was brought about; and this will take us through the year 1834." Similarly, what was originally an organic pattern of meaningful events and images evoking ships and the sea as metaphors of Douglass's self becomes in later versions mere literary allusions, as in the following: "My poor weather-beaten bark now reached smoother water and gentler breezes. My stormy life at Covey's had been of service to me. The things that would have seemed very hard had I gone directly to Mr. Freeland's from the home of Master Thomas, were now 'trifles light as air.'"¹

On the other hand, Douglass never fitted himself to popular stereotypes of the ex-slave as did Josiah Henson, nor did the later autobiographies always lapse into chatty, meandering memoirs. During the darkest times of Reconstruction, as his accounts attest, Douglass bore the banner of black independence, insisting on the freedman's rights to "the ballot-box, the jury-box, and the cartridge-box." When temporarily barred from the White House at Lincoln's second inaugural reception, he commented wryly but gently of the servants: "They were simply complying with an old custom, the outgrowth of slavery, as dogs will sometimes rub their necks, long after their collars are removed, thinking they are still there." In general, however, he was unable to sustain a sharp sense of his own voice and identity through a long historical narrative. After reading the *Narrative* and then turning to *Life and Times*, one has difficulty agreeing with Rayford Logan in calling the later autobiography a "classic." This reader records a different impression—of the imaginative unity and superior force of the young man's self and story. Reading all three versions of this remarkable life makes one recognize afresh the difficulties of dealing with the long sweep of a public and private history. Douglass is no more to be criticized for writing more than one (and more than one kind of) personal history than are W. E. B. Du Bois or Mark Twain or Gertrude Stein. But if *Life and Times of Frederick Douglass* shows some of the strains of multiple autobiography, the *Narrative* should remind us how hard it is to repeat an early success. But then many of the most compelling black autobiographies have been the work of the young—*Black Boy, The Autobiography of Malcolm X, Manchild in the Promised Land, I Know Why the Caged Bird Sings.* Their precursor is the *Narrative of the Life of Frederick Douglass, an American Slave, Written by Himself.* It is the first native American autobiography to create a black identity in a style and form adequate to the pressures of historic black experience.

Animal Farm Unbound

H. Bruce Franklin

Narrative of the Life of Frederick Douglass, an American Slave, Written by Himself is a book created by a being who was once considered an animal, even by himself, for an audience that remains unconvinced that he is in fact a fellow human being. So it should come as no surprise that animal imagery embodies Douglass's deepest meanings.

In *Long Black Song,* Houston Baker, Jr., notes that animal metaphors "appear in most of the chapters of the *Narrative.*" Baker offers several explanations. He observes that "Douglass is aware of American slavery's chattel principle, which equated slaves with livestock, and he is not reluctant to employ animal metaphors to capture the general inhumanity of the system." He makes an intriguing suggestion about overtones in the *Narrative* from the animal tales of black slave culture. And he emphasizes the appropriativeness of the animal imagery to "the agrarian settings and characters." Albert Stone disagrees with Baker, arguing that Douglass's "images of ships and the sea" are far more central than animal imagery, forming a pattern which "connects and defines all stages of his personal history." Stone's sensitive exploration of the nautical imagery is a valuable contribution to our appreciation of the artistic richness of this book. It is, however, the animal imagery that is crucial, and in ways far more significant than even Baker perceived.

These images not only structure the development of the *Narra-*

From *New Letters* 43, no. 3 (April 1977). ©1977 by the Curators of the University of Missouri, Kansas City.

tive, but also locate the book on the front lines of a major ideological battleground of the 1840s and 1850s. Douglass is asking, and answering, one central question in the *Narrative:* What is a human being? That is, within his historical context, how is a human being different from animals (or machines) that can perform labor? This was also the central philosophical and scientific question of his time, a question that all our subsequent history has been trying to resolve. While Douglass wrote, Darwin and Marx were both wrestling with precisely the same question. And in America, natural science and its definition of what was human was in the process of coming to focus most narrowly on "the Negro."

Slavery, as we now recognize, went through a fundamental change around 1830, completing its evolution from a predominantly small-scale, quasi-domestic institution appended to handtool farming and manufacture into the productive base of an expanding agricultural economy, utilizing machinery to process the harvested crops and pouring vast quantities of agricultural raw materials, principally cotton, into developing capitalist industry in the northern states and England. Prior to the 1830s, as George Fredrickson documents in *The Black Image in the White Mind,* open assertions of the "*permanent* inferiority" of blacks "were exceedingly rare." In fact, many eighteenth-century apologists for slavery defended it as a means of "raising" and "civilizing" the poor, benighted, childlike Negro. But in the 1830s there emerged in America a world view based on the belief that blacks were inherently a race inferior to whites, and as part of this world view there developed a scientific theory of blacks as beings half way, or even less than half way, between animals and white people. This was part of the shift of blacks from their role as children, appropriate to a professedly patriarchal society which offered them the means of eventual development into adulthood, into their role as subhuman beasts of burden, the permanent mainstay of the labor force of expanding agribusiness.

By 1833, this world view had been scientifically formulated in Richard Colfax's *Evidence against the Views of the Abolitionists, Consisting of Physical and Moral Proofs of the Natural Inferiority of the Negroes.* In his researches into the skulls and facial angles of Negroes, Colfax prefigured the developed science of the 1840s and 1850s known as the "American School of Ethnology." He argued that "the acknowledged meanness of the Negro's intellect only coincides with the shape of his head." This can be readily seen in the Negro's "facial angle," which was "almost to a level with that of a brute." Colfax concludes that Negroes are half way between animals and white people: "the Negroes, whether physically or morally considered, are so inferior as to resemble the brute creation as nearly as they do the white species." (Fredrickson cites this and many other works prior to 1845 making the same biological case against the Negro.)

Colfax did not further develop the concept of Negroes as a distinct *species,* but by the late 1830s this next logical position was achieving its first systematic presentation in a body of scientific literature dedicated to demonstrating "that the black man was a member of a separate and permanently inferior species." In the early 1840s came the theory of polygenesis. Dr. Samuel George Morton proved scientifically in *Crania Americana* and *Crania Aegyptiaca* that Negroes did not descend from Adam but were a distinct and subhuman species originating in southern Africa. Carolyn Karcher has shown how Melville satirized this "science" in her "Melville's 'The 'Gees': A Forgotten Satire on Scientific Racism."

Frederick Douglass had lived the social reality which these scientific theories were adduced to perpetuate. He had begun life as a farm animal. Looking back, he traces the course of his development into a conscious human being, threatened all along the way by the danger of being reduced once again to a beast. Using the most brilliant manipulation of his audience's literary conventions to display the particularities of his own experience, Douglass is able to show what it means to be a human being in an age and society dominated by racist ideology and maintaining its basic productive activities through the use of one class of human beings as work animals by another class of human beings. For Douglass, as for Karl Marx, writing the previous year in what we now call the *Economic and Philosphic Manuscripts of 1844,* human beings are distinguished as a species by a creative consciousness which derives from the circumstances of their existence; this consciousness gives us the potential freedom to change those circumstances to meet human needs and desires, and it is in the struggle for that freedom that this consciousness develops.

The first paragraph of the *Narrative of the Life of Frederick Douglass, an American Slave, Written by Himself* is concerned with the basic circumstances of his birth—place and date. Douglass has no problem locating the place and he does so, in the first sentence, establishing at once the artfully restrained, almost unemotional, matter-of-fact style which is to be the underlying norm for the entire narrative: "I was born in Tuckahoe, near Hillsborough, and about twelve miles from Easton, in Talbot county, Maryland." But the second sentence poses a problem for this precise, no-nonsense narrator: "I have no accurate knowledge of my age, never having seen any authentic record containing it." In dryly explaining his predicament to the reader, Douglass can only compare himself and his fellow slaves to other farm animals: "slaves know as little of their age as horses know of theirs." This is the starting point of his consciousness, something like a human, something like a beast.

Like most slaves Douglass never knew his father. He learns, however, that his father was a white man, quite possibly his master, one of those who made the satisfaction of his "lusts" both "profitable as well as pleasurable" by increasing the number of his slaves. So Douglass himself apparently was created through the sexual union of the two "species" of beings defined by those scientists of the 1840s, and one of these—the loftier—would probably gain a profit from the transaction when the little suckling became marketable. Following the "common custom," his mother is deliberately separated from him while he is a small baby: "I never saw my mother, to know her as such, more than four or five times in my life; and each of these times was very short in duration, and at night."

The little boy's first consciousness of the meaning of slavery comes through the spectacle of his beautiful aunt being whipped by his master, apparently because of sexual jealousy. The master "stripped her from neck to waist," tied her hands to an overhead hook, and then proceeded to "whip upon her naked back till she was literally covered with blood":

> The louder she screamed, the harder he whipped; and where the blood ran fastest, there he whipped longest. He would whip her to make her scream, and whip her to make her hush; and not until overcome by fatigue, would he cease to swing the blood-clotted cowskin. I remember the first time I ever witnessed this horrible exhibition. I was quite a child, but I well remember it. I never shall forget it whilst I remember any thing.

The words the master uses over and over again to define Douglass's aunt while he flagellates her cannot be repeated to the polite readers of the *Narrative*. Douglass has to record them as "'you d—d b—h.'" But their meaning is clear enough, for they signify the essence of the slaveowners' views of their black slaves. The human master is merely punishing a female animal.

As for the little boy, he was but "seldom whipped," as Douglass tells us in a passage that I believe stands as one of the most brilliant achievements in style and content of nineteenth-century American prose:

> I was seldom whipped by my old master, and suffered little from any thing else than hunger and cold. I suffered much from hunger, but much more from cold. In hottest summer and coldest winter, I was kept almost naked—no shoes, no stockings, no jacket, no trousers, nothing on but a coarse tow linen shirt, reaching only to my knees. I had no bed. I must have perished with cold, but that, the coldest nights, I used to steal a bag which was used for carrying corn to the mill. I would crawl into this bag, and there sleep on the cold, damp, clay floor, with my head in and feet out. My feet have been so cracked with the frost, that the pen with which I am writing might be laid in the gashes.

After the first two sentences, simple but meticulously balanced, the style becomes stripped and stark, almost as naked as the little boy it describes living, or rather existing, on the level of brute survival. On the surface almost laconic, the passage virtually explodes with artfully arranged, highly volatile tensions. The first great disparity is between the little boy and the man writing his story, who is the little boy grown up. The two worlds in which they live are brought into direct physical contact as the writer takes his pen and lays it in the frost-cracked gashes on the boy's feet. By using the tool with which he is communicating to his polite audience as the implement of yoking in these two worlds, he also forces that audience to join him in contacting the boy. And in that conjunction, he brings his readers face to face with the first of many moral inversions: to survive, the slave must violate the property rights defined by society; he must steal a bag intended to help produce profit. In all this, we are forced to sense a tremendous disparity between the emotional level of the prose, running on that matter-of-fact norm, and the potential rage and violence implicit in the slave's situation. This is all part of Douglass's patient preparation for the climax of his *Narrative,* and for his final warning to his audience.

Douglass next describes how he ate: "our food was coarse corn meal boiled. This was called *mush*. It was put into a large wooden tray or trough, and set down upon the ground. The children were then called, like so many pigs, and like so many pigs they would come and devour the mush." In the very next paragraph, Douglass tells of his leaving this plantation. He thus establishes the juxtaposition which will provide one underlying dialectic for the rest of the narrative, the dialectic between rural and urban existence. Here we see most clearly an opposition of values between Douglass's vision, which is generally representative of his black contemporaries, and the vision dominant in most of the white literature of the period.

The movement from country to city, and the conflict between the values of these two worlds, was of course a highly conventional literary theme in antebellum America, with its rapid industrialization and urbanization. This is, most typically, envisioned as a fall from rural innocence and natural freedom into the artificialities of the infernal city, as in Hawthorne's "My Kinsman, Major Molineux" and Melville's *Pierre; Or, The Ambiguities.* Outside the city is the Eden to which the conscious person may wish to return, rarely with as much success as in the visions projected by Thoreau.

For Frederick Douglass, the movement primarily means the opposite. The city to him represents consciousness and the possibility of freedom; the country represents brutalization and the certainty of slavery. So the boy, now "probably between seven and eight years old," spends almost three days "in the creek, washing off the plantation scruff," "for the people in Baltimore were very cleanly, and would laugh at me if I looked dirty." He is going to be given a pair of trousers; "The thought of owning a pair of trousers was great indeed!" To merit the trousers and the city, he must no longer be a young pig: "It was almost a sufficient motive, not only to make me take off what would be called by pig-drovers the mange, but the skin itself."

In the city, Douglass becomes a house boy. His expectations about life in the city are not disappointed: "A city slave," he discovers, "is almost a freeman, compared with a slave on the plantation." And there he encounters simultaneously two great sources of knowledge. The first, introduced by his mistress, is the alphabet. The second, confronting him in the form of his master's reactions, is that all the values of the slave must be the opposite of those of the slave-owner. The master forbids his wife from any further instruction of the boy because "it was unlawful, as well as unsafe, to teach a slave to read":

> To use his own words, further, he said, "If you give a nigger an inch, he will take an ell. A nigger should know nothing but to obey his master—to do as he is told to do. Learning would *spoil* the best nigger in the world. Now," said he, "if you teach that nigger (speaking of myself) how to read, there would be no keeping him. It would forever unfit him to be a slave."

So the master points to consciousness as the means to freedom, to the written language as a means to increase consciousness, and to himself as the negation of consciousness, the negation that must constantly be negated in order to achieve freedom:

> These words sank deep into my heart, stirred up sentiments within that lay slumbering, and called into existence an entirely new train of thought. . . . I now understood what had been to me a most perplexing difficulty—to wit, the white man's power to enslave the black man. It was a grand achievement, and I prized it highly. From that moment, I understood the pathway from slavery to freedom. . . . The very decided manner with which he spoke, and strove to impress his wife with the evil consequences of giving me instruction, served to convince me that he was deeply sensible of the truths that he was uttering. It gave me the best assurance that I might rely with the utmost confidence on the results which, he said, would flow from teaching me to read. What he most dreaded, that I most desired. What he most loved, that I most hated. That which to him was a great evil, to be carefully shunned, was to me a great good, to be diligently sought; and the argument which he so warmly urged, against my learning to read, only served to inspire me with a desire and determination to learn.

This experience, which foreshadows the climax of the *Narrative,* defines for Douglass the path to consciousness and freedom, that is to humanity. Unlike Pinocchio, who can become human only by learning to be honest, Douglass can attain his humanity only by learning deceit and trickery. He reveals to us some of the wily and devious tricks he uses, still as a small boy, to gain from the hostile white world around him the ability to read and write. Frankness, trustfulness, humility, passivity are all for him just so many snares that would put him back in the barnyard with the horses and pigs.

Douglass succeeds in learning how to read, and the master's worst fears come true:

> The more I read, the more I was led to abhor and detest my enslavers. I could regard them in no other light than a band of successful robbers, who had left their homes, and gone to Africa, and stolen us from our homes, and in a strange land reduced us to slavery.

But as his eyes are opened, as he gains intensifying consciousness of his own condition, without seeing how to change it, his own transformation becomes the source of his greatest torment. He now sometimes yearns to be deprived of consciousness, to be, in fact, an unthinking animal: "I have often wished myself a beast. I preferred the condition of the meanest reptile to my own. Any thing, no matter what, to get rid of thinking!"

In the following chapter his urban sanctuary is disrupted by a temporary fall back into the barnyard. The death of his legal owner forces him back to be present at the redivision of all the property. Although now only about ten or eleven years old, he understands the scene all too well. It is perhaps the most conventional scene in the slave narrative genre, undoubtedly because it was such a critical event in the actual lives of the slaves and one that displayed most dramatically the essence of chattel slavery. It is when the slaves are "divided, like so many sheep." For Douglass the young city slave it is a revelation, which Douglass the author presents as the literal embodiment of his animal imagery:

> We were all ranked together at the valuation. Men and women, old and young, married and single, were ranked with horses, sheep, and swine. There were horses and men, cattle and women, pigs and children, all holding the same rank in the scale of being, and were all subjected to the same narrow examination. Silvery-headed age and

sprightly youth, maids and matrons, had to undergo the same indelicate inspection.

He concludes this paragraph by foreshadowing a reversal that will take place in the function of the animal imagery: "At this moment, I saw more clearly than ever the brutalizing effects of slavery upon both slave and slaveholder."

Earlier, Douglass had traced the degradation of his mistress, who "at first lacked the depravity indispensable to shutting me up in mental darkness." In order to become "equal to the task of treating me as though I were a brute," she must be transformed from being a "tender-hearted woman" to a creature of "tiger-like fierceness." As the *Narrative* progresses, Douglass makes us increasingly aware of this other kind of animal. Just as the slaves in the early part of the book are likened to barnyard animals, the slaveowners later are compared more and more to predatory beasts. So when Douglass is lucky enough to be returned temporarily to Baltimore, the fate he thus escapes is "worse than lion's jaws." But at the age of about fourteen, this is just the fate he meets, as he is returned to the plantation.

His new country master cannot tame him, even with "a number of severe whippings." So he decides to rent young Douglass out to Edward Covey, a farmer notorious as a "'nigger-breaker.'" Douglass now finds himself, "for the first time in my life, a field hand." It is now 1833, the very year in which Richard Colfax was publishing his evidence that "the Negroes, whether physically or morally considered, are so inferior as to resemble the brute creation as nearly as they do the white species."

Edward Covey is known to his slaves, significantly enough, as "the snake." Vicious as he is, Covey's main weapon in breaking slaves is not the whip but work. At the very moment that Colfax is propagandizing the concept that Negroes are inherently and permanently subhuman, Douglass is learning that through unending work a person can be transformed into a beast:

> We were worked in all weathers. It was never too hot or too cold; it could never rain, blow, hail, or snow, too hard for us to work in the field. Work, work, work, was scarcely more the order of the day than of the night. The longest days were too short for him, and the shortest nights too long for him. I was somewhat unmanageable when I first went there, but a few months of this discipline

> tamed me. Mr. Covey succeeded in breaking me. I was broken in body, soul, and spirit. My natural elasticity was crushed, my intellect languished, the disposition to read departed, the cheerful spark that lingered about my eye died; the dark night of slavery closed in upon me; and behold a man transformed into a brute!
>
> Sunday was my only leisure time. I spent this in a sort of beast-like stupor, between sleep and wake, under some large tree. At times I would rise up, a flash of energetic freedom would dart through my soul, accompanied with a faint beam of hope, that flickered for a moment, and then vanished.

This is the lowest point in Douglass's life, and its essential crisis. Reduced to animal existence, his human consciousness seems to serve only the function of self torture. But even this ultimate degradation contains the potential of human liberation.

Douglass's authorial strategy here is crucial. He is aware that his audience has been conditioned to think of him as half human. He does not protest by proclaiming that he is every bit as human as the reader. Instead, he takes the reader through his own experience of becoming, in fact, "beast-like," and not through extraordinary or exceptional torture but through unremitting, mindless labor without end, the ordinary life of the slave. By conceding that he himself had become like an animal *after* attaining a much higher consciousness, Douglass forces the reader to recognize that he or she, merely by being cast down from his or her relatively comfortable social existence, could also be reduced to the semblance of an animal. This experience is quite different from the one in which Douglass the child had first awakened to find himself a slave. How could Douglass's readers possibly imagine themselves as beings who had never known any existence but that as a rural beast of burden? The Douglass who returns to the animal farm is much closer to the typical reader: he has lived in the city; he has thought philosophically about freedom and slavery; he can read and write; he has read books. Thus Douglass can serve as a surrogate for the reader, and the reader may be able to share a portion of Douglass's slave experience. The readers can discover that all their book knowledge and philosophical consciousness would not serve to distinguish *them* from animals if they were suddenly plunged into plantation slavery. The situation resembles those in many science fiction stories, from *Gulliver's Travels* and Voltaire's *Micromegas* to *Planet of the Apes* and James McConnell's "Learning Theory," in which human beings find themselves incapable of demonstrating to alien creatures that they are part of an intelligent species.

Douglass, however, does find the way to demonstrate—to Covey, to himself, and thus to the readers—that he is a human being. This is the key event in his life, the climax of his narrative, and the core of his philosophical, historical, and practical message. Frederick Douglass now speaks in the second person, addressing the reader directly: "You have seen how a man was made a slave; you shall see how a slave was made a man."

Douglass discovers that it is not cranial capacity or facial angles or book knowledge or intelligence in the abstract that distinguishes the human species from brutes. It is the consciousness which allows people to alter the conditions of existence, a consciousness that develops in the struggle for freedom from brute necessity. Faced with slavery, that can mean only one thing before all: "I resolved to fight; and, suiting my action to the resolution, I seized Covey hard by the throat . . . I told him . . . that he had used me like a brute for six months, and that I was determined to be used so no longer." They fight for what seems hours. Douglass overcomes Covey, "the snake." Then comes the famous passage which all students of the *Narrative* have seen as its heart and climax:

> This battle with Mr. Covey was the turning-point in my career as a slave. It rekindled the few expiring embers of freedom, and revived within me a sense of my own manhood. It recalled the departed self-confidence, and inspired me again with a determination to be free. The gratification afforded by the triumph was a full compensation for whatever else might follow, even death itself. He only can understand the deep satisfaction which I experienced, who has himself repelled by force the bloody arm of slavery. I felt as I never felt before. It was a glorious resurrection, from the tomb of slavery, to the heaven of freedom. My long-crushed spirit rose, cowardice departed, bold defiance took its place; and I now resolved that, however long I might remain a slave in form, the day had passed forever when I could be a slave in fact.

This is a troublesome passage for many professors of literature, for it challenges their most fundamental assumptions about the relations between body and mind, and between life and art. I noted earlier that no journal except one devoted to Afro-American literature had ever found an article on the *Narrative* acceptable for publication. There may be many possible reasons for this, but one that is in hand reveals what is fundamentally at issue in the rejection of Douglass's art and vision. In recommending rejection of an article on the *Narrative* submitted to an academic journal, a referee insisted that the author made a serious error in not finding any "irony in the situation in which Douglass must reduce his conflict with the slaveholders to a question of brute strength and physical violence in order to assert his 'manhood.'" This referee went on to explain the values the author of the article *ought* to have shared to have his view of the *Narrative* acceptable:

> Does he [Douglass] learn that in matching the brute in himself against the animalism of his enslavery that he becomes the victor? If [the author of the article] is right about Douglass' genius, it would seem more convincing that Douglass recognized that his real victory over slavery and his most splendid assertion of his manhood was the *Narrative* itself: his triumph over language and his own rage.

Characteristic of his social class, this academic referee equates the body and physical violence, no matter how it is exerted, with "the brute," and the mind, especially evidenced in its verbal products, with what is really human. This is based on the underlying academic dichotomy between mind and body, an expression of bourgeois ideology, which envisions workers as mindless bodies and intellectuals as pure minds whose bodily physical comforts have nothing to do with their thinking. In "On the Teaching of Literature in the Highest Academies of the Empire," I showed how this dichotomy structures the most fundamental unexamined assumptions governing the study and teaching of literature in America. The primary of these assumptions I caricatured in these terms:

> First, there is the overall relationship between art and life. Great literary art transcends life. That is, literary achievements are more significant than social or political actions.

Or rather, I thought this was a caricature until I saw that statement by the anonymous referee about the climax of Douglass's *Narrative*.

Douglass's individual rebellion, his personal repelling "by force the bloody arm of slavery," has tremendous importance for him, for the history of nineteenth-century America, for us. As Nancy T. Clasby has shown, "Douglass' act of violent resistance and the mysterious rebirth he experienced" are "crucial thematic elements" not just in this *Narrative* but in black literature up through the present. As Clasby perceives:

> The institutions under which Douglass had lived had failed to give him a viable identity—his manhood. The fight with Covey symbolically shattered the institutions and the old identity.

To be reborn as a human being, to shed his animal identity imposed upon him by the white man, this black slave must commit the most forbidden crime of all: he must strike the white man who oppresses him.

Not to understand the meaning of this is to fail to comprehend not only Douglass's *Narrative* but the historical epoch we ourselves live in, an epoch characterized by the anticolonial struggles of the nonwhite peoples of the world. Frantz Fanon, the black psychiatrist and revolutionary theorist, has written several books unfolding the historical implications of the psychological truths Douglass was able to compress into a paragraph. As Fanon puts it, the act of violence against the oppressor, even on the individual level, is the primal event that "frees the native from his inferiority complex and from his despair and inaction; it makes him fearless and restores his self-respect" (*The Wretched of the Earth*). For Douglass, as for the peoples studied by Fanon, the initial act of violence is the premise of a new community for the oppressed. As Clasby argues, Douglass's entire subsequent life as a leader of his people flowed from this act:

> From the time of his resolution that "the day had passed forever" when he could be "a slave in fact," Douglass experienced his own integrity, a love for his brothers, and a relationship to spiritual realities which had been denied him by the conventional societal mechanisms. For the family which had been denied him by slavery he found a new brotherhood among his fellow rebels. He traded

"slaveholding Christianity" for a close-knit and loving community of suffering slaves.

Douglass's creation of this *Narrative* is also a monumental act, but it was contingent upon what he did that day on the plantation. And the brilliant art of this narrative embodies in animal imagery his rebirth into a new identity. Unlike those professors who think a person becomes a "brute" when he or she fights back against oppression, Douglass has shown us that the brutes on the farm remain sheepish and that it is human beings who can learn how to resist and defeat slavery. Prior to this, as we have seen, Douglass compares himself and other slaves to those domesticated farm animals—horses, pigs, sheep— and compares the slaveowners and their accomplices to wild predators—lions, tigers, snakes. From this point on in the *Narrative,* Douglass never again likens himself or any salve to an animal. The animal imagery associated with the slaveholders, however, continues, actually building to a climax after his escape from both enslavement and the rural world. This climax takes place in the least agrarian setting—New York City. The main animal in this vision is a crocodile, not a mammal but a reptile, not American, but African. The jungle fantasies of the American ethnologists, facial angles and all, are brought back to their origin.

"Immediately after my arrival in New York," he tells us, "I felt like one who had escaped a den of hungry lions." But this feeling "very soon subsided," as he realizes that this great metropolis of America is part of a "hunting-ground for slaveholders—whose inhabitants are legalized kidnappers." He becomes aware that he is "every moment subjected to the terrible liability of being seized upon by his fellowmen, as the hideous crocodile seizes upon his prey!" He is frighteningly alone in this urban jungle, "among fellowmen, yet feeling as if in the midst of wild beasts": "I was afraid to speak to any one for fear of speaking to the wrong one, and thereby falling into the hands of money-loving kidnappers, whose business it was to lie in wait for the panting fugitive, as the ferocious beasts of the forest lie in wait for their prey." So Douglass turns his readers' world upside down. They may still wonder if he is really a human being like themselves or just some lower species in human clothing. He knows that *he* is human, and he warns them of what *they* will be if they collaborate with the crocodiles and other beasts whose laws govern America.

Frederick Douglass was about twenty-seven years old when he published the *Narrative,* his first book, in 1845. The following year another twenty-seven-year old American author, Herman Melville, published, in England, his first book, *Narrative of a Four Months' Residence among the Natives of a Valley of the Marquesas Islands; Or, A Peep at Polynesian Life.* Douglass, writing as a nonwhite slave in white America, had to veil some of his message in imagery. Melville, writing as a white American who had lived in a nonwhite society under the shadow of imperialism, spoke more bluntly when he distinguished "the white civilized man as the most ferocious animal on the face of the earth." When Melville's *Narrative* was published in America as *Typee,* these words, along with many other crucial passages, were deleted. When Douglass's *Narrative* was published in America, he had to flee his native land. His owner, backed up by the laws of the United States of America, was seeking to hunt down and recapture his runaway beast of burden, the author of *Narrative of the Life of Frederick Douglass, an American Slave, Written by Himself.*

Narration, Authentication, and Authorial Control

Robert B. Stepto

The strident, moral voice of the former slave recounting, exposing, appealing, apostrophizing, and above all, *remembering* his ordeal in bondage is the single most impressive feature of a slave narrative. This voice is striking not only because of what it relates but because the slave's acquisition of that voice is quite possibly his only permanent achievement once he escapes and casts himself upon a new and larger landscape. In their most elementary form, slave narratives are, however, full of other voices that are frequently just as responsible for articulating a narrative's tale and strategy. These other voices may be those of various "characters" in the "story," but mainly they are those found in the appended documents written by slaveholders and abolitionists alike. These documents—and voices—may not always be smoothly integrated with the former slave's tale, but they are nevertheless parts of the narrative. Their primary function is, of course, to authenticate the former slave's account; in doing so, they are at least partially responsible for the narratives being accepted as historical evidence. However, in literary terms, the documents collectively create something close to a dialogue—of forms as well as of voices—which suggests that in its primal state or first phase the slave narrative is an eclectic narrative form.

When the various forms (letters, prefaces, guarantees, tales) and

---

From *Afro-American Literature: The Reconstruction of Instruction,* edited by Dexter Fisher and Robert B. Stepto. ©1978 by the Modern Language Association of America.

their accompanying voices become integrated in the slave narrative text, we are presented with another type of basic narrative which I call an integrated narrative. This type of narrative represents the second phase of narration in the slave narrative and usually yields a more sophisticated text, wherein most of the literary and rhetorical functions previously performed by several texts and voices (the appended prefaces, letters, and documents as well as the tale) are now rendered by a loosely unified single text and voice. In this second phase, the authenticating documents "come alive" in the former slave's tale as speech and even action; and the former slave—often while assuming a deferential posture toward his white friends, editors, and guarantors—carries much of the burden of introducing and authenticating his own tale. In short, a second-phase narrative is a more sophisticated narrative because the former slave's voice assumes many more responsibilities than that of recounting the tale.

Because an integrated or second-phase narrative is less a collection of texts and more a unified narrative, we may say that, in terms of narration, the integrated narrative is in the process of becoming—irrespective of authorial intent—a generic narrative, by which I mean a narrative of discernible genre such as history, fiction, essay, or autobiography. This process is no simple "gourd vine" activity: An integrated narrative does not become a generic narrative "overnight," and, indeed, there are no assurances that in becoming a new type of narrative it is transformed automatically into a distinctive generic text. What we discover, then, is a third phase to slave narrative narration wherein two developments may occur: The integrated narrative (Phase II) is dominated either by its tale or by its authenticating strategies. In the first instance, the narrative and moral energies of the former slave's voice and tale so resolutely dominate those of the narrative's authenticating machinery (voices, documents, rhetorical strategies) that the narrative becomes in thrust and purpose far more metaphorical than rhetorical. When the integrated narrative becomes in this way a figurative account of action, landscape, and heroic self-transformation, it is so close generically to history, fiction, and autobiography that I term it a generic narrative.

In the second instance, the authenticating machinery either remains as important as the tale or actually becomes, usually for some purpose residing outside the text, the dominant and motivating feature of the narrative. Since this is also a sophisticated narrative phase, figurative presentations of action, landscape, and self may also occur, but such developments are rare and always ancillary to the central thrust of the text. When the authenticating machinery is dominant in this fashion, the integrated narrative becomes an authenticating narrative.

As these remarks suggest, one reason for investigating the phases of slave narrative narration is to gain a clearer view of how some slave narrative types become generic narratives and how, in turn, generic narratives—once formed, shaped, and set in motion by certain distinctly Afro-American cultural imperatives—have roots in the slave narratives. This bears as well on our ability to distinguish between narrative modes and forms and to describe what we see. When, for example, a historian or literary critic calls a slave narrative an autobiography, what he *sees* is, most likely, a narrative told in the first person that possesses literary features distinguishing it from the ordinary documents providing historical and sociological data. But a slave narrative is not necessarily an autobiography. We need to know the finer shades between the more easily discernible categories of narration, and we must discover whether these stops arrange themselves in progressive, contrapuntal, or dialectic fashion—or whether they possess any arrangement at all. As the scheme described above and diagrammed below suggests, I believe there are at least four identifiable modes of narration within the slave narrative, all of which have a direct bearing on the development of subsequent Afro-American narrative forms.

Phase I: basic narrative (a): "eclectic narrative"—authenticating documents and strategies (sometimes including one by the author of the tale) *appended* to the tale

↓

Phase II: basic narrative (b): "integrated narrative"—authenticating documents and strategies integrated into the tale and formally becoming voices and/or characters in the tale

Phase III:

(a) "generic narrative"—authenticating documents and strategies are totally

(b) "authenticating narrative"—the tale is subsumed by the authenticating strat-

subsumed by the tale; the slave narrative becomes an identifiable generic text, e.g., autobiography, etc.

egy; the slave narrative becomes an authenticating document for other, usually generic, texts, e.g., novel, history

II

What we observe in the first two phases of slave narrative narration is the former slave's ultimate lack of control over his own narrative occasioned primarly by the demands of audience and authentication. This dilemma is not unique to the authors of these narratives; indeed, many modern black writers still do not control their personal history once it assumes literary form. For this reason, Frederick Douglass's *Narrative of the Life of Frederick Douglass, an American Slave, Written by Himself* (1845) seems all the more a remarkable literary achievement. Because it contains several segregated narrative texts—a preface, a prefatory letter, the tale, an appendix—it appears to be, in terms of the narrative phases, a rather primitive slave narrative. But each of the ancillary texts seems to be drawn to the tale by some sort of extraordinary gravitational pull or magnetic attraction. There is, in short, a dynamic energy between the tale and each supporting text; the Douglass narrative is an integrated narrative of a very special order. While the integrating process does, in a small way, pursue the conventional path of creating characters out of authenticating texts (William Lloyd Garrison silently enters Douglass's tale at the very end), its new and major thrust is the creation of that aforementioned energy that binds the supporting texts to the tale while at the same time removing them from participation in the narrative's rhetorical and authenticating strategies. In short, Douglass's tale dominates the narrative and does so because it alone authenticates the narrative.

The introductory texts to the tale are two in number: a preface by William Lloyd Garrison, the famous abolitionist and editor of *The Liberator;* and a "Letter from Wendell Phillips, Esq.," who was equally renowned as an abolitionist, a crusading lawyer, and a judge. In theory, each of these introductory documents should be a classic guarantee written almost exclusively to a white reading public, concerned primarily and ritualistically with the white validation of a newfound black voice, and removed from the tale in such ways that the guarantee and tale vie silently and surreptitiously for control of the narrative as a whole. But these entries simply are not fashioned that way. To be sure, Garrison offers a conventional guarantee when he writes:

> Mr. DOUGLASS has very properly chosen to write his own Narrative, in his own style, and according to the best of his ability, rather than to employ some one else. It is, therefore, entirely his own production; and . . . it is, in my judgment, highly creditable to his head and heart.

And Phillips, while addressing Douglass, most certainly offers a guarantee to "another" audience as well:

> Every one who has heard you speak has felt, and, I am confident, every one who reads your book will feel, persuaded that you give them a fair specimen of the whole truth. No one-sided portait,—no wholesale complaints,— but strict justice done, whenever individual kindliness has neutralized, for a moment, the deadly system with which it was strangely allied.

But these passages dominate neither the tone nor the substance of their respective texts.

Garrison is far more interested in writing history (specifically that of the 1841 Nantucket Anti-Slavery Convention and the launching of Douglass's career as a lecture agent for various antislavery societies) and recording his own place in it. His declaration, "I shall never forget his [Douglass's] first speech at the convention," is followed shortly thereafter by "*I rose,* and declared that Patrick Henry of revolutionary fame, never made a speech more eloquent in the case of liberty. . . . *I reminded* the audience of the peril which surrounded this self-emancipated young man. . . . *I appealed* to them, whether they would ever allow him to be carried back into slavery,— law or no law, constitution or no constitution" (italics added). His preface ends, not with a reference to Douglass or to his tale, but with an apostrophe very much like one he would use to exhort and arouse an antislavery assembly. In short, with the following cry, Garrison hardly guarantees Douglass's tale but reenacts his own abolitionist career instead:

Reader! are you with the man-stealers in sympathy and purpose, or on the side of their down-trodden victims? If with the former, then are you the foe of God and man. If with the latter, what are you prepared to do and dare in their behalf? Be faithful, be vigilant, be untiring in your efforts to break every yoke, and let the oppressed go free. Come what may—cost what may—inscribe on the banner which you unfurl to the breeze, as your religious and political motto—NO COMPROMISE WITH SLAVERY! NO UNION WITH SLAVEHOLDERS!

In the light of this closure and (no matter how hard we try to ignore it) the friction that developed between Garrison and Douglass in later years, we might be tempted to see Garrison's preface at war with Douglass's tale for authorial control of the narrative as a whole. Certainly, there is a tension, but that tension is stunted by Garrison's enthusiasm for Douglass's tale:

> This *Narrative* contains many affecting incidents, many *passages* of great eloquence and power; but I think the most thrilling one of them all is the *description* DOUGLASS gives of his feelings, as he stood soliloquizing respecting his fate, and the chances of his one day being a free man.... Who can read that *passage,* and be insensible to its pathos and sublimity? (italics added)

What Garrison does, probably subconsciously, is an unusual and extraordinary thing—he becomes the first guarantor we have seen who not only directs the reader to the tale but also acknowledges the tale's singular rhetorical power. Thus, Garrison enters the tale by being at the Nantucket Convention with Douglass in 1841 and by authenticating the impact of the tale, not its facts. He fashions his own apostrophe, but finally he remains a member of Douglass's audience far more than he assumes the posture of a competing or superior voice. In this way, Garrison's preface stands outside Douglass's tale but is steadfastly bound to it.

This is even more so the case for Wendell Phillips's "Letter." It contains passages that seem to be addressed to credulous readers in need of a "visible" authority's guarantee, but by and large the "Letter" is directed to Frederick Douglass alone. It opens with "My Dear Friend," and there are many extraliterary reasons for wondering initially if the friend is actually Frederick. Shortly thereafter, however, Phillips declares, "I am glad the time has come when the 'lions write history,'" and it becomes clear that he not only addresses Douglass but also writes in response to the tale. These features, plus Phillips's specific references to how Douglass acquired his "ABC" and learned of "where the 'white sails' of the Chesapeake were bound," serve to integrate Phillips's "Letter" into Douglass's tale. Above all, we must see in what terms the "Letter" is a cultural and linguistic event: Like the Garrison document, it presents it author as a member of Douglass's audience, but the act of letterwriting, of correspondence, implies a moral and linguistic parity between a white guarantor and black author that we have not seen before and that we do not always see in American literary history *after* 1845. In short, the tone and posture initiated in Garrison's preface are completed and confirmed in Phillips's "Letter," and while these documents are integrated into Douglass's tale, they remain segregated outside the tale in the all-important sense that they yield Douglass sufficient narrative and rhetorical space in which to render personal history in—and as—a literary form.

What marks Douglass's narration and control of his tale is his extraordinary ability to pursue several types of writing with ease and with a degree of simultaneity. The principal types of writing we discover in the tale are syncretic phrasing, introspective analysis, internalized documentation, and participant-observation. Of course, each of these types has its accompanying authorial posture, the result being that even the telling of the tale (as distinct from the content of the tale) yields a portrait of a complex individual marvelously facile in the tones, shapes, and dimensions of his voice.

Douglass's syncretic phrasing is often discussed, and the passage most widely quoted is probably "My feet have been so cracked with the frost, that the pen with which I am writing might be laid in the gashes." The remarkable clarity of this language needs no commentary, but what one admires as well is Douglass's ability to startlingly conjoin past and present and to do so with images that not only stand for different periods in his personal history but also, in their fusion, speak of his evolution from slavery to freedom. The pen, symbolizing the quest for literacy fulfilled, actually takes measure of the wounds of the past, and this measuring process becomes a metaphor in and of itself for the artful composition of travail transcended. While I admire this passage, the syncretic phrases I find even more intriguing are those that pursue a kind of acrid punning upon the names of Douglass's oppressors. A minor example appears early in the tale, when Douglass deftly sums up an overseer's character by writing, "Mr. Severe was rightly named: he was a cruel man." Here, Douglass is content with "glossing" the name; but late in the tale, just before attempting to escape in 1835, Douglass takes another oppressor's name and does not so much gloss it or play with it as *work upon* it to such an extent that, riddled with irony, it is devoid of its original meaning:

> At the close of the year 1834, Mr. Freeland again hired me of my master, for the year 1835. But, by this time, I began to want to live *upon free land* as well as *with Freeland;* and I was no longer content, therefore, to live with him or any other slaveholder.

Of course, this is effective writing—far more effective than what is found in the average slave narrative—but the point I wish to make is that Douglass seems to fashion these passages for both his readership and himself. Each example of his wit and increasing facility with language charts his ever-shortening path to literacy; thus, in their way, Douglass's syncretic phrases reveal his emerging comprehension of freedom and literacy and are another introspective tool by which he may benchmark his personal history.

But the celebrated passages of introspective analysis are even more pithy and direct. In these, Douglass fashions language as finely honed and balanced as an aphorism or Popean couplet, and thereby orders his personal history with neat, distinct, and credible moments of transition. When Mr. Auld forbids Mrs. Auld to instruct Douglass in the ABC, for example, Douglass relates:

> From that moment, I understood the pathway from slavery to freedom.... Whilst I was saddened by the thought of losing the aid of my kind mistress, I was gladdened by the invaluable instruction which, by the merest accident, I gained from my master.

The clarity of Douglass's revelation is as unmistakable as it is remarkable. As rhetoric, the passage is successful because its nearly extravagant beginning is finally rendered quite acceptable by the masterly balance and internal rhyming of "saddened" and "gladdened," which is persuasive because it is pleasant and because it offers the illusion of a reasoned conclusion.

Balance is an important feature of two other equally celebrated passages that quite significantly open and close Douglass's telling of his relations with Mr. Covey, an odd (because he worked in the fields alongside the slaves) but vicious overseer. At the beginning of the episode, in which Douglass finally fights back and draws Covey's blood, he writes:

> You have seen how a man was made a slave; you shall see how a slave was made a man.

And at the end of the episode, to bring matters linguistically and narratively full circle, Douglass declares:

> I now resolved that, however long I might remain a slave in form, the day has passed forever when I could be a slave in fact. I did not hesitate to let it be known of me, that the white man who expected to succeed in whipping, must also succeed in killing me.

The sheer poetry of these statements is not lost on us, nor is the fact of why the poetry was created in the first place. One might suppose that in another age Douglass's determination and rage might take a more effusive expression, but I cannot imagine that to be the case. In the first place, his linguistic model is obviously scriptural; and in the second, his goal is the presentation of a historical self, not the record of temporary hysteria. This latter point persuades me that Douglass is about the business of discovering how personal history may be transformed into autobiography. Douglass's passages of introspective analysis almost single-handedly create fresh space for themselves in the American literary canon.

Instead of reproducing letters and other documents written by white guarantors within the tale or transforming guarantors into characters, Douglass internalizes documents that, like the syncretic and introspective passages, order his personal history. For example, Douglass's discussion of slave songs begins with phrases such as "wild songs" and "unmeaning jargon" but concludes, quite typically for him, with a study of how he grew to "hear" the songs and how the hearing affords yet another illumination of his path from slavery to freedom:

> I did not, when a slave, understand the deep meaning of those rude and apparently incoherent songs. I was myself within the circle; so that I neither saw nor heard as those without might see and hear. They told a tale of woe which was then altogether beyond my feeble comprehension. . . . Every tone was a testimony against slavery, and a prayer to God for deliverance from chains. The hearing of those wild notes always depressed my spirit, and filled me with ineffable sadness. I have frequently found myself in tears while hearing them. The mere recurrence to those songs, even now, afflicts me; and while I am writing these lines, an expression of feeling has already found its way down my cheek.

The tears of the past and present interflow, and Douglass not only documents his saga of enslavement but also, with typical recourse to syncretic phrasing and introspective analysis, advances his presentation of self.

Douglass's other internalized documents are employed with comparable efficiency as we see in the episode where he attempts an escape in 1835. In this episode, the document reproduced is the pass or "protection" Douglass wrote for himself and his compatriots in the escape plan:

> "This is to certify that I, the undersigned, have given the bearer, my servant, full liberty to go to Baltimore, and spend the Easter holidays. Written with mine own hand, &c., 1835.
> "WILLIAM HAMILTON,
> "Near St. Michael's, in Talbot county, Maryland."

The protection exhibits Douglass's increasingly refined sense of how to manipulate language—he has indeed come a long way from that day Mr. Auld halted his ABC lessons—but even more impressive, I believe, is the act of reproducing the document itself. We know from the tale that when their scheme was thwarted, each slave managed to destroy his pass, so Douglass reproduces his language from memory, and there is no reason to doubt a single jot of his recollection. My point here is simply that Douglass can draw so easily from the wellsprings of memory because the protection is not a mere scrap of memorabilia but rather a veritable road sign on his path to freedom and literacy. In this sense, his protection assumes a place in Afro-American letters as a key antecedent to such documents as the fast-yellowing notes of James Weldon Johnson's Ex-Coloured Man and "The Voodoo of Hell's Half Acre" in Richard Wright's *Black Boy.*

All of the types of narrative discourse discussed thus far reveal features of Douglass's particular posture as a participant-observer narrator. But the syncretic phrases, introspective studies, and internalized documents only exhibit Douglass as a teller and doer, and part of the great effect of his tale depends upon what Douglass does not tell, what he refuses to reenact in print. Late in the tale, at the beginning of chapter 11, Douglass writes:

> I now come to that part of my life during which I planned, and finally succeeded in making, my escape from slavery. But before narrating any of the peculiar circumstances, I deem it proper to make known my intention not to state all the facts connected with the transaction. . . . I deeply regret the necessity that impels me to suppress any thing of importance connected with my experience in slavery. It would afford me great pleasure indeed, as well as materially add to the interest of my narrative, were I at liberty to gratify a curiosity, which I know exists. . . . But I must deprive myself of this pleasure, and the curious gratification which such a statement would afford. I would allow myself to suffer under the greatest imputations which evil-minded men might suggest, rather than exculpate myself, and thereby run the hazard of closing the slightest avenue by which a brother slave might clear himself of the chains and fetters of slavery.

It has been argued that one way to test a slave narrative's authenticity is by gauging how much space the narrator gives to relating his escape as opposed to describing the conditions of his captivity. If the adventure, excitement, and perils of the escape seem to be the raison d'être for the narrative's composition, then the narrative is quite possibly an exceedingly adulterated slave's tale or a bald fiction. The theory does not always work perfectly: Henry "Box" Brown's narrative and that of William and Ellen Craft are predominantly recollections of extraordinary escapes, and yet, as far as we can tell, these are authentic tales. But the theory nevertheless has great merit, and I have often wondered to what extent it derives from the example of

Douglass's tale and emotionally, if not absolutely rationally, from his fulminations against those authors who unwittingly excavate the underground railroad and expose it to the morally thin mid-nineteenth-century American air. Douglass's tale is spectacularly free of suspicion, because he never tells a detail of his escape to New York, and it is this marvelously rhetorical omission or silence that both sophisticates and authenticates his posture as a participant-observer narrator. When a narrator wrests this kind of preeminent authorial control from the ancillary voices "circling" his narrative, we may say that he controls the presentation of his personal history and that his tale is becoming autobiographical. In this light the last few sentences of Douglass's tale take on special meaning:

> But, while attending an anti-slavery convention at Nantucket, on the 11th of August, 1841, I felt strongly moved to speak.... It was a severe cross, and I took it up reluctantly. The truth was, I felt myself a slave, and the idea of speaking to white people weighed me down. I spoke but a few moments, when I felt a degree of freedom, and said what I desired with considerable ease. From that time until now, I have been engaged in pleading the case of my brethren—with what success, and what devotion, I leave those acquainted with my labors to decide.

With these words, the narrative, as many have remarked, comes full circle, taking us back, not to the beginning of the *tale,* but rather to Garrison's prefatory remarks on the Convention and Douglass's first public address. This return may be pleasing in terms of the sense of symmetry it affords, but it is also a remarkable feat of rhetorical strategy: Having traveled with Douglass through his account of his life, we arrive in Nantucket in 1841 to hear him speak and, in effect, to become, along with Mr. Garrison, his audience. The final effect is that Douglass reinforces his posture as an articulate hero while supplanting Garrison as the definitive historian of his past.

Even more important, I think, is the final image Douglass bestows of a slave shedding his last fetter and becoming a man by first finding his voice and then, as sure as light follows dawn, speaking "with considerable ease." In one brilliant stroke, the quest for freedom and literacy implied from the start even by the narrative's title is resolutely consummated.

The final text of the narrative, the appendix, is a discourse by

Douglass on his view of Christianity and Christian practice as opposed to what he exposed in his tale to be the bankrupt, immoral faith of slaveholders. As rhetorical strategy, the discourse is effective generally because it lends weight and substance to what passes for a conventional complaint of slave narrative narrators and because Douglass's exhibition of faith can only enhance his already considerable posture as an articulate hero. But more specifically, the discourse is most efficacious because at its heart lies a vitriolic poem written by a Northern Methodist minister, which Douglass introduces by writing,

> I conclude these remarks by copying the following portrait of the religion of the south, (which is, by communion and fellowship, the religion of the north,) which I soberly affirm is "true to life," and without caricature or the slightest exaggeration.

The poem is strong and imbued with considerable irony, but what we must appreciate here is the effect of the white Northerner's poem conjoined with Douglass's authentication of the poem. The tables are clearly reversed. Douglass has controlled his personal history and at the same time fulfilled the prophecy suggested in his implicit authentication of Garrison's preface: He has explicitly authenticated what is conventionally a white Northerner's validating text. Douglass's narrative thus offers what is unquestionably our best portrait in Afro-American letters of the requisite act of assuming authorial control. An author can go no further than Douglass did without writing all the texts constituting the narrative himself.

Binary Oppositions in Chapter One of the *Narrative*

Henry Louis Gates, Jr.

> *I was not hunting for my liberty, but also hunting for my name.*
> WILLIAM WELLS BROWN, 1849

> *Whatever may be the ill or favored condition of the slave in the matter of mere personal treatment, it is the chattel relation that robs him of his manhood.*
> JAMES PENNINGTON, 1849

> *When at last in a race a new principle appears, an idea,—that conserves it; ideas only save races. If the black man is feeble and not important to the existing races, not on a parity with the best race, the black man must serve, and be exterminated. But if the black man carries in his bosom an indispensable element of a new and coming civilization; for the sake of that element, no wrong nor strength nor circumstance can hurt him: he will survive and play his part. . . . I esteem the occasion of this jubilee to be the proud discovery that the black race can contend with the white: that in the great anthem which we call history, a piece of many parts and vast compass, after playing a long time a very low and subdued accompaniment, they perceive the time arrived when they can strike in with effect and take a master's part in the music.*
> EMERSON, 1844

> *The white race will only respect those who oppose their usurpation, and acknowledge as equals those who will not submit to their rule. . . . We must make an issue, create an event and establish for ourselves a position. This is essentially necessary for our effective elevation as a people, directing our destiny and redeeming ourselves as a race.*
> MARTIN R. DELANY, 1854

From *Afro-American Literature: The Reconstruction of Instruction*, edited by Dexter Fisher and Robert B. Stepto. © 1978 by the Modern Language Association of America.

Autobiographical forms in English and in French assumed narrative priority toward the end of the eighteenth century; they shaped themselves principally around military exploits, court intrigues, and spiritual quests. As Stephen Butterfield has outlined, "Elizabethan sea dogs and generals of the War of the Spanish Succession wrote of strenuous campaigns, grand strategy, and gory battles. The memoirs of Louis XIV's great commander, the Prince of Condé, for example, thrilled thousands in Europe and America, as did the 'inside stories' of the nefarious, clandestine doings of the great European courts. The memoirs of the Cardinal De Retz, which told the Machiavellian intrigues of French government during Louis XIV's minority and of the cabal behind the election of a pope, captivated a large audience. Even more titillating were personal accounts of the boudoir escapades of noblemen and their mistresses. Nell Gwyn, Madame Pompadour, and even the fictitious Fanny Hill were legends if not idols in their day. More edifying but no less marvelous were the autobiographies of spiritual pilgrimage—such as the graphic accounts of Loyola, John Bunyan, and the Quaker George Fox. Their mystical experiences and miraculous deliverances filled readers with awe and wonder." It is no surprise, then, that the narratives of the escaped slave became, during the three decades before the Civil War, the most popular form of written discourse in the country. Its audience was built to order. And the expectations created by this peculiar autobiographical convention, as well as by two other literary traditions, had a profound effect on the shape of discourse in the slave narrative. I am thinking here of the marked (but generally unheralded) tradition of the sentimental novel and, more especially, of the particularly American transmutation of the European picaresque. The slave narrative, I suggest, is a "countergenre," a mediation between the novel of sentiment and the picaresque, oscillating somewhere between the two in a bipolar moment, set in motion by the mode of the Confession. (Indeed, as we shall see, the slave narrative spawned its formal negation, the plantation novel.)

Claudio Guillén's seminal typology of the picaresque, outlined as seven "characteristics" of that form and derived from numerous examples in Spanish and French literature, provides a curious counterpoint to the morphology of the slave narratives and aids remarkably in delineating what has proved to be an elusive, but recurring, narrative structure.

The picaro, who is after all a type of character, only becomes one at a certain point in his career, just as a man or woman "becomes" a slave only at a certain (and structurally crucial) point of perception in his or her "career." Both the picaro and the slave narrators are orphans; both, in fact, are outsiders. The picaresque is a pseudo-autobiography, whereas the slave narratives often tend toward quasi-autobiography. Yet in both, "life is at the same time revived and judged, presented and remembered." In both forms, the narrator's point of view is partial and prejudiced, although the total view "of both is reflective, philosophical, and critical on moral or religious grounds." In both, there is a general stress on the material level of existence or indeed of *subsistence,* such as sordid facts, hunger, and money. There is in the narration of both a profusion of objects and detail. Both the picaro and the slave, as outsiders, comment on if not parody collective social institutions. Moreover, both, in their odysseys, move horizontally through space and vertically through society.

If we combine these resemblances with certain characteristics of the sentimental novel, such as florid asides, stilted rhetoric, severe piety, melodramatic conversation, destruction of the family unit, violation of womanhood, abuse of innocence, punishment of assertion, and the rags-to-riches success story, we can see that the slave narrative grafted together the conventions of two separate literary traditions and became its own form, utilizing popular conventions to affect its reader in much the same way as did cheap, popular fiction. Lydia Child, we recall, was not only the amanuensis for the escaped slave, Harriet Jacobs, but also a successful author in the sentimental tradition. (That the plantation novel was the antithesis or negation of the slave narrative becomes apparent when we consider its conventions. From 1824, when George Tucker published *The Valley of the Shenandoah,* the plantation novel concerned itself with aristocratic, virtuous masters; beastlike, docile slaves; great manor houses; squalid field quarters; and idealized, alabaster womanhood—all obvious negations of themes common to the slave narratives. Indeed, within two years of the publication in 1852 of Harriet Beecher Stowe's *Uncle Tom's Cabin,* at least fourteen plantation novels appeared.)

It should not surprise us, then, that the narratives were popular, since the use of well-established and well-received narrative conventions was meant to ensure commercial and hence political success. By at least one account, the sale of the slave narratives reached such profound proportions that a critic [George R. Graham] was moved to complain that the "shelves of booksellers groan under the weight of Sambo's woes, done up in covers! . . . We hate this niggerism, and hope it may be done away with. . . . If we are threatened with any more negro stories—here goes." These "literary nigritudes" [sic], as he calls them, were "stories" whose "editions run to hundreds of thousands." Marion Wilson Starling recalls Gladstone's belief that not more than about five percent of the books published in England had a sale of more than five hundred copies; between 1835 and 1863, no fewer than ten of these were slave narratives. So popular were they in England that a considerable number were published at London or Manchester before they were published in America, if at all. Nor should it surprise us that of these, the more popular were those that defined the genre structurally. It was Frederick Douglass's *Narrative* of 1845 that exploited the potential of and came to determine the shape of language in the slave narrative.

Douglass's *Narrative,* in its initial edition of five thousand copies, was sold out in four months. Within a year, four more editions of two thousand copies each were published. In the British Isles, five editions appeared, two in Ireland in 1846 and three in England in 1846 and 1847. Within the five years after its appearance, a total of some thirty thousand copies of the *Narrative* had been published in the English-speaking world. By 1848, a French edition, a paperback, was being sold in the stalls. *Littells Living Age,* an American periodical, gave an estimate of its sweep in the British Isles after one year's circulation: "Taking all together, not less than one million persons in Great Britain and Ireland have been excited by the book and its commentators."

Of the scores of reviews of the *Narrative,* two, especially, discuss the work in terms of its literary merits. One review, published initially in the *New York Tribune* and reprinted in *The Liberator,* attempts to place the work in the larger tradition of the narrative tale as a literary form.

> Considered merely as a narrative, we have never read one more simple, true, coherent, and warm with genuine feeling. It is an excellent piece of writing, and on that score to be prized as a specimen of the powers of the black race, which prejudice persists in disputing. We prize highly all evidence of this kind, and it is becoming more abundant.

Even more telling is the review from the *Lynn Pioneer* reprinted in the same issue of *The Liberator;* this review was perhaps the first to attempt to attach a priority to the *Narrative*'s form and thereby place Douglass directly in a major literary tradition.

> It is evidently drawn with a nice eye, and the coloring is chaste and subdued, rather than extravagant or overwrought. Thrilling as it is, and full of the most burning eloquence, it is yet simple and unimpassioned.

Although its "eloquence is the eloquence of truth," and so "is as simple and touching as the impulses of childhood," yet its "message" transcends even its superior moral content: "There are passages in it which would brighten the reputation of any author,—while the book, as a whole, judged as a mere work of art, would widen the fame of Bunyan or De Foe." Leaving the matter of "truth" to the historians, these reviews argue correctly that despite the intention of the author for his autobiography to be a major document in the abolitionist struggle and regardless of Douglass's meticulous attempt at documentation, the *Narrative* falls into the larger class of the heroic fugitive with some important modifications that are related to the confession and the picaresque forms (hence, Bunyan and Defoe), a peculiar blend that would mark Afro-American fiction at least from the publication of James Weldon Johnson's *Autobiography of an Ex-Coloured Man.*

These resemblances between confession and picaresque informed the narrative shape of Afro-American fiction in much the same way as they did in the English and American novel. As Robert Scholes and Robert Kellogg maintain,

> The similarity in narrative stance between picaresque and confession enables the two to blend easily, making possible an entirely fictional narrative which is more in the spirit of the confession than the picaresque, such as *Moll Flanders* and *Great Expectations.*

But the same blend makes possible a different sort of sublime narrative, "one that is *picaresque* in spirit but which employs actual materials from the author's life, such as [Wells's] *Tono-Bungay.*" Into this class fall slave narratives, the polemical Afro-American first-person form the influence of which would shape the development of point of view in black fiction for the next one hundred years, precisely because

> by turning the direction of the narrative inward the author almost inevitably presents a central character who is an example of something. By turning the direction of the narrative outward the author almost inevitably exposes weaknesses in society. First-person narrative is thus a ready vehicle for ideas.

It is this first-person narration, utilized precisely in this manner, that is the first great shaping characteristic of the slave narratives. But there is another formal influence on the slave narratives the effect of which is telling: this is the American romance.

Like Herman Melville's marvelous romance, *Pierre,* the slave narratives utilize as a structural principle the irony of seeming innocence. Here in American society, both say, is to be found as much that is contrary to moral order as could be found in prerevolutionary Europe. The novelty of American innocence is, however, the refusal or failure to recognize evil while participating in that evil. As with other American romantic modes of narration, the language of the slave narratives remains primarily an expression of the self, a conduit for particularly personal emotion. In this sort of narrative, language was meant to be a necessary but unfortunate instrument merely. In the slave narratives, this structuring of the self couples with the minute explication of gross evil and human depravity, and does so with such sheer intent as to make for a tyranny of point. If the matter of the shaping of the self can come only after the slave is free, in the context of an autobiographical narrative where he first posits that full self, then slavery indeed dehumanizes and must in no uncertain terms be abolished, by violence if necessary, since it is by nature a violent institution. The irony here is tyrannically romantic: illusion and substance are patterned antitheses.

As with other examples of romance, the narratives turn on an unconsummated love: The slave and the ex-slave are the dark ladies of the new country destined to expire for unrequited love. Yet the leitmotif of the journey north and the concomitant evolution of consciousness within the slave—from an identity as property and object to a sublime identity as a human being and subject—display in the first person the selfsame spirit of the New World's personal experience with Titanic nature that Franklin's *Autobiography* has come to symbolize. The author of the slave narrative, in his flight through the wilderness (re-created in vivid detailed descriptions of the relation between man and land on the plantation and off), seems to be arguing strongly that man can "study nature" to know himself. The two great precepts—the former Emersonian and the latter Cartesian—in the American adventure become one. Further, as with the American symbolists, the odyssey is a process of *becoming*: Whitman, for instance, is less concerned with explorations of emotion than with exploration as a mode of consciousness. Slave narratives not only describe the voyage but also enact the voyage so that their content is primarily a reflection of their literary method. Theirs is a structure in which the writer and his subject merge into the stream of language. Language indeed is primarily a perception of reality. Yet, unlike the American symbolists, these writers of slave narratives want not so much to adopt a novel stance from which the world assumes new shapes as to impose a new form onto the world. There can be no qualification as to the nature of slavery; there can be no equivocation.

Stephen Butterfield explicates this idea rather well by contrasting the levels of diction in the slave narrative *The Life of John Thompson* with a remarkably similar passage from Herman Melville's *Moby-Dick*.

The first is from Thompson:

> The harpoon is sharp, and barbed at one end, so that when it has once entered the animal, it is difficult to draw it out again, and has attached to its other end a pole, two inches thick and five feet long. Attached to this is a line 75 to 100 fathoms in length, which is coiled into the bow of the boat.

Melville follows:

> Thus the whale-line folds the whole boat in its complicated coils, twisting and writhing about it in almost every direction. All the oarsmen are involved in its perilous contortions; so that to the timid eye of the landsman they seem as festooning their limbs.

There is a difference here of rhetorical strategies that distinguishes the two. Melville's language is symbolic and weighted with ambiguous moral meanings: The serpentine rope allows for no innocence;

"all the oarsmen" are involved, even those who have nothing to do with coiling it in the tub; the crew lives with the serpent and by the serpent, necessarily for their livelihood, unaware of the nature of the coil yet contaminated and imperiled by its inherent danger. Melville thus depicts the metaphysical necessity of evil.

John Thompson's language is distinguished formally from the concrete and symbolistic devices in Melville. Thompson allows the imagery of a whaling voyage to carry moral and allegorical meanings, yet he means his narration to be descriptive and realistic; his concern is with verisimilitude. There can be nothing morally ambiguous about the need to abolish slavery, and there can be little ambiguity about the reason for the suffering of the slave. "The slave narrative," Butterfield concludes, "does not see oppression in terms of a symbol-structure that transforms evil into a metaphysical necessity. For to do so would have been to locate the source of evil outside the master-slave relationship, and thus would have cut the ideological ground from under the entire thrust of the abolitionist movement." Thompson means not so much to narrate as to convey a message, a value system; as with the black sermon, the slave's narrative functions as a single sign. And the nature of Frederick Douglass's rhetorical strategy directly reflects this sentiment through the use of what rhetoricians have called antitheses and of what the structuralists have come to call the binary opposition.

In the act of interpretation, we establish a sign relationship between the description and a meaning. The relations most crucial to structural analysis are functional binary oppositions. Roman Jakobson and Morris Halle argue in *Fundamentals of Language* that binary oppositions are inherent in all languages, that they are, indeed, a fundamental principle of language formation itself. Many structuralists, seizing on Jakobson's formulation, hold the binary opposition to be a fundamental operation of the human mind, basic to the production of meaning. Levi-Strauss, who turned topsy-turvy the way we examine mythological discourse, describes the binary opposition as "this elementary logic which is the smallest common denominator of all thought." Levi-Strauss's model of opposition and mediation, which sees the binary opposition as an underlying structural pattern as well as a method for revealing that pattern, has in its many variants become a most satisfying mechanism for retrieving almost primal social contradictions, long ago "resolved" in the mediated structure itself. Perhaps it is not irresponsible or premature to call Levi-Strauss's contribution to human understanding a classic one.

Frederic Jameson, in *The Prison-House of Language,* maintains that

> the binary opposition is . . . at the outset a heuristic principle, that instrument of analysis on which the mythological hermeneutic is founded. We would ourselves be tempted to describe it as a technique for stimulating perception, when faced with a mass of apparently homogeneous data to which the mind and the eyes are numb: a way of forcing ourselves to perceive difference and identity in a wholly new language the very sounds of which we cannot yet distinguish from each other. It is a decoding or deciphering device, or alternately a technique of language learning.

How does this "decoding device" work as a tool to practical criticism? When any two terms are set in opposition to each other the reader is forced to explore qualitative similarities and differences, to make some connection, and, therefore, to derive some meaning from points of disjunction. If one opposes A to B, for instance, and X to Y, the two cases become similar as long as each involves the presence and absence of a given feature. In short, two terms are brought together by some quality that they share and are then opposed and made to signify the absence and presence of that quality. The relation between presence and absence, positive and negative signs, is the simplest form of the binary opposition. These relations, Jameson concludes, "embody a tension 'in which one of the two terms of a binary opposition is apprehended as positively having a certain feature while the other is apprehended as deprived of the feature in question.' "

Frederick Douglass's *Narrative* attempts with painstaking verisimilitude to reproduce a system of signs that we have come to call plantation culture, from the initial paragraph of chapter 1:

> I was born in Tuckahoe, near Hillsborough, and about twelve miles from Easton, in Talbot County, Maryland. I have no accurate knowledge of my age, never having seen any authentic record containing it. By far the larger part of the slaves know as little of their ages as horses know of theirs, and it is the wish of most masters within my knowledge to keep their slaves thus ignorant. I do not remember to have ever met a slave who could tell of his

> birthday, they seldom come nearer to it than planting-time, harvest-time, cherry-time, spring-time, or fall-time. A want of information concerning my own was a source of unhappiness to me even during childhood. The white children could tell their ages. I could not tell why I ought to be deprived of the same privilege. I was not allowed to make any inquiries of my master concerning it. He deemed such inquiries on the part of a slave improper and impertinent, and evidence of a restless spirit. The nearest estimate I can give makes me now between twenty-seven and twenty-eight years of age. I come to this, from hearing my master say, some time during 1835, I was about seventeen years old.

We see an ordering of the world based on a profoundly relational type of thinking, in which a strict barrier of difference or opposition forms the basis of a class rather than, as in other classification schemes, an ordering based on resemblances or the identity of two or more elements. In the text, we can say that these binary oppositions produce through separation the most inflexible of barriers: that of meaning. We, the readers, must exploit the oppositions and give them a place in a larger symbolic structure. Douglass's narrative strategy seems to be this: He brings together two terms in special relationships suggested by some quality that they share; then, by opposing two seemingly unrelated elements, such as the sheep, cattle, or horses on the plantation and the specimen of life known as slave, Douglass's language is made to signify the presence and absence of some quality—in this case, humanity. Douglass uses this device to explicate the slave's understanding of himself and of his relation to the world through the system of the perceptions that defined the world the planters made. Not only does his *Narrative* come to concern itself with two diametrically opposed notions of genesis, origins, and meaning itself, but its structure actually turns on an opposition between nature and culture as well. Finally and, for our purposes, crucially, Douglass's method of complex mediation—and the ironic reversals so peculiar to his text—suggests overwhelmingly the completely arbitrary relation between description and meaning, between signifier and signified, between sign and referent.

Douglass uses these oppositions to create a unity on a symbolic level, not only through physical opposition but also through an opposition of space and time. The *Narrative* begins "I was born in Tuckahoe, near Hillsborough, and about twelve miles from Easton, in Talbot County, Maryland." Douglass knows the physical circumstances of his birth: Tuckahoe, we know, is near Hillsborough and is twelve miles from Easton. Though his place of birth is fairly definite, his date of birth is not for him to know: "I have no accurate knowledge of my age," he admits, because "any authentic record containing it" would be in the possession of others. Indeed, this opposition, or counterpoint, between that which is *knowable* in this world of the slave and that which is *not,* abounds throughout this chapter. Already we know that the world of the master and the world of the slave are separated by an inflexible barrier of meaning. The knowledge the slave has of his circumstances he must deduce from the *earth;* a quantity such as time, our understanding of which is *cultural* and not *natural,* derives from a nonmaterial source, let us say the *heavens:* "The white children could tell their ages. I could not."

The deprivation of the means to tell the time is the very structural center of this initial paragraph: "A want of information concerning my own [birthday] was a source of unhappiness to me even during childhood." This state of disequilibrium motivates the slave's search for his humanity as well as Douglass's search for his text. This deprivation has created that gap in the slave's imagination between self and other, between black and white. What is more, it has apparently created a relation of likeness between the slave and the animals. "By far," Douglass confesses, "the large part of slaves know as little of their ages as horses know of theirs." This deprivation is not accidental; it is systematic: "it is the wish of most masters within my knowledge to keep their slaves thus ignorant." Douglass, in his subtle juxtaposition here of "masters" and "knowledge" and of "slaves" and "ignorance," again introduces homologous terms. "I do not remember to have ever met a slave," Douglass emphasizes, "who could tell of his birthday." Slaves, he seems to conclude, are they who cannot plot their course by the linear progression of the calendar. Here, Douglass summarizes the symbolic code of this world, which makes the slave's closest blood relations the horses and which makes his very notion of time a cyclical one, diametrically opposed to the master's linear conception: "They [the slaves] seldom come nearer to [the notion of time] than planting-time, harvest-time, cherry-time, spring-time, or fall-time." The slave had arrived, but not *in time* to partake at the welcome table of human culture.

For Douglass, the bonds of blood kinship are the primary metaphors of human culture. As an animal would know its mother, so Douglass knows his. "My mother was named Harriet Bailey. She was the daughter of Isaac and Betsey Bailey." Both of whom were "colored," Douglass notes, "and quite dark." His mother "was of a darker complexion" even than either grandparent. His father, on the other hand, is some indefinite "white man," suggested through innuendo to be his master: "The opinion was also whispered," he says, "that my master was my father." His master was his father; his father his master: "of the correctness of this opinion," Douglass concludes, "I know nothing," only and precisely because "the means of knowing was withheld from me." Two paragraphs below, having reflected on the death of his mother, Douglass repeats this peculiar unity twice again. "Called thus suddenly away," he commences, "she left me without the slightest intimation of who my father was." Yet Douglass repeats "the whisper that my father was my master" as he launches into a description of the rank odiousness of a system "that slaveholders have ordained, and by law established," in which the patrilinear succession of the planter has been forcibly replaced by a matrilinear succession for the slave: "the children of slave women shall in all cases follow the condition of their mothers." The planters therefore make of the "gratification of their wicked desires," spits Douglass, a thing "profitable as well as pleasurable." Further, the end result of "this cunning arrangement" is that "the slaveholder, in cases not a few, sustains to his slaves the double relation of master and father." "I know of such cases," he opens his sixth paragraph, using a declaration of verisimilitude as a transition to introduce another opposition, this one between the fertile slave-lover-mother and the planter's barren wife.

The profound ambiguity of this relationship between father and son and master and slave persists, if only because the two terms "father" and "master" are here embodied in one, with no mediation between them. It is a rather grotesque bond that links Douglass to his parent, a bond that embodies "the distorted and unnatural relationship endemic to slavery." It is as if the usually implied primal tension between father and son is rendered apparent in the daily contact between father-master-human and son-slave-animal, a contact that occurs, significantly, only during the light of day.

Douglass's contact with his mother ("to know her as such," he qualifies) never occurred "more than four or five times in my life."

Each of these visits, he recalls, "was of short duration," and each, he repeats over and over, took place "at night." Douglass continues: "[My mother] made her journey to see me in the night, travelling the whole distance," he mentions as if an afterthought, "on foot." "I do not recollect of ever seeing my mother," he repeats one sentence later, "by the light of day. She was with me in the *night*" (emphasis added). Always she returned to a Mr. Stewart's plantation, some twelve miles away, "long before I waked" so as to be at the plantation before dawn, since she "was a field hand, and a whipping is the penalty of not being in the field at sunrise." The slaves, metaphorically, "owned" the night, while the master owned the day. By the fourth paragraph of the narrative, the terms of our homology—the symbolic code of this world—are developed further to include relations of the animal, the mother, the slave, the night, the earth, matrilinear succession, and nature opposed to relations of the human being, the father, the master, the daylight, the heavens, patrilinear succession, and culture. Douglass, in short, opposes the absolute and the eternal to the mortal and the finite. Our list, certainly, could be expanded to include oppositions between spiritual/material, aristocratic/base, civilized/barbaric, sterile/fertile, enterprise/sloth, force/principle, fact/imagination, linear/cyclical, thinking/feeling, rational/irrational, chivalry/cowardice, grade/brutishness, pure/cursed, and human/beastly.

Yet the code, Douglass proceeds to show, stands in defiance of the natural *and* moral order. Here Douglass commences as mediator and as trickster to reverse the relations of the opposition. That the relation between the slave-son and his master-father was an unnatural one and even grotesque, as are the results of any defilement of Order, is reflected in the nature of the relation between the plantation mistress and the planter's illegitimate offspring. "She is ever disposed to find fault with them," laments Douglass; "she is never better pleased than when she sees them under the lash." Indeed, it is the white mistress who often compels her husband, the master, to sell "this class of his slaves, out of deference to the feelings of his white wife." But it is the priority of the economic relation over the kinship tie that is the true perversion of nature in this world: "It is often the dictate of humanity for a man to sell his own children to human flesh-mongers," Douglass observes tellingly. Here we see the ultimate reversal: For it is now the mistress, the proverbial carrier of culture, who demands that the master's son be delivered up to the "human flesh-mongers" and traded for consumption. Douglass has here defined American cannibalism, a consumption of human flesh dictated by a system that could only be demonic.

Douglass's narrative demonstrates not only how the deprivation of the hallmarks of identity can affect the slave but also how the slaveowner's world negates and even perverts those very values on which it is built. Deprivation of a birth date, a name, a family structure, and legal rights makes of the deprived a brute, a subhuman, says Douglass, until he comes to a consciousness of these relations; yet it is the human depriver who is the actual barbarian, structuring his existence on the consumption of human flesh. Just as the mulatto son is a mediation between two opposed terms, man and animal, so too has Douglass's text become the complex mediator between the world as the master would have it and the world as the slave knows it really is. Douglass has subverted the terms of the code he was meant to mediate: He has been a trickster. As with all mediations the trickster is a mediator and his mediation is a trick—only a trick; for there can be no mediation in this world. Douglass's narrative has aimed to destroy that symbolic code that created the false oppositions themselves. The oppositions, all along, were only arbitrary, not fixed.

Douglass first suggests that the symbolic code created in this text is arbitrary and not fixed, human-imposed not divinely ordained in an ironic aside on the myth of the curse of Ham, which comes in the very center of the seventh paragraph of the narrative and which is meant to be an elaboration on the ramifications of "this class of slaves" who are the fruit of the unnatural liaison between animal and man. If the justification of this order is the curse on Ham and his tribe, if Ham's tribe signified the black African, and if this prescription for enslavement is scriptural, then, Douglass argues, "it is certain that slavery at the south must soon become unscriptural; for thousands are ushered into the world, annually, who, like myself, owe their existence to white fathers, and those fathers," he repeats for the fourth time, are "most frequently their own masters."

As if to underscore the falsity of this notion of an imposed, inflexibly divine order, Douglass inverts a standard Christian symbol, that of the straight and narrow gate to Paradise. The severe beating of his Aunt Hester, who "happened," Douglass advises us parenthetically, "to be absent when my master desired her presence," is the occasion of this inversion. "It struck me with awful force," he remembers. "It was the blood-stained gate," he writes, "the entrance to the hell of slavery, through which I was about to pass. It was," he concludes, "a most terrible spectacle." This startling image suggests that of the archetypal necromancer, Faustus, in whose final vision the usual serene presence of the Cross is stained with warm and dripping blood.

Douglass has posited the completely arbitrary nature of the sign. The master's actions belie the metaphysical suppositions on which is based the order of this world: It is an order ostensibly imposed by the Father of Adam, yet one in fact exposed by the sons of Ham. It is a world the oppositions of which have generated their own mediator, Douglass himself. This mulatto son, half-animal, half-man, writes a text (which is itself another mediation) in which he can expose the arbitrary nature of the signs found in this world, the very process necessary to the destruction of this world. "You have seen how a man was made a slave," Douglass writes at the structural center of his *Narrative*, "you shall see how a slave was made a man." As with all mediation, Douglass has constructed a system of perception that becomes the plot development in the text but that results in an inversion of the initial state of the oppositions through the operations of the mediator himself as indicated in this diagram:

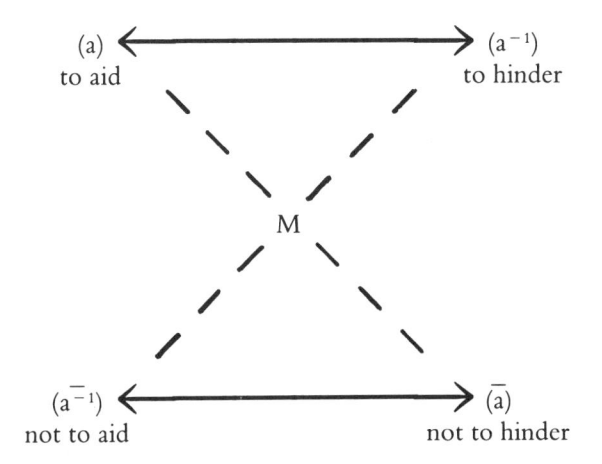

$$
\begin{array}{ccc}
(a) & \longleftrightarrow & (a^{-1}) \\
\text{to aid} & & \text{to hinder} \\
 & M & \\
(\overline{a^{-1}}) & \longleftrightarrow & (\overline{a}) \\
\text{not to aid} & & \text{not to hinder}
\end{array}
$$

With this narrative gesture alone, slave has become master, creature has become man, object has become subject. What more telling embodiment of Emersonian idealism and its "capacity" to transubstantiate a material reality! Not only has an *idea* made subject of object, but creature has assumed self and the assumption of self has created a race. For, as with all myths of origins, the relation of self to race is a relation of synecdoche. As Michael Cooke maintains concerning the characteristics of black autobiography:

> The self is the source of the system of which it is a part, creates what it discovers, and although (as Coleridge realized) it is nothing unto itself, it is the possibility of everything for itself. Autobiography is the coordination of the self as content—everything available in memory, perception, understanding, imagination, desire—and the self as shaped, formed in terms of a perspective and pattern of interpretation.

If we step outside the self-imposed confines of chapter 1 to seek textual evidence, the case becomes even stronger. The opposition between culture and nature is clearly contained in a description of a slave meal, found in chapter 5 [of *Afro-American Literature*]. "We were not regularly allowanced. Our food was coarse corn meal boiled. This was called *mush.* It was put into a large wooden tray or trough, and set down upon the ground. The children were then called, like so many pigs, and like so many pigs they would come and devour the mush; some with oyster-shells, others with pieces of shingle, some with naked hands, and none with spoons. He that ate fastest got most; he that was strongest secured the best place; and few left the trough satisfied." The slave, we read, did not eat food; he ate mush. He did not eat with a spoon; he ate with pieces of shingle, or on oyster shells, or with his naked hands. Again we see the obvious culture-nature opposition at play. When the slave, in another place, accepts the comparison with and identity of a "bad sheep," he again has inverted the terms, supplied as always by the master, so that the unfavorable meaning that this has for the master is supplanted by the favorable meaning it has for the slave. There is in this world the planter has made, Douglass maintains, an ironic relation between appearance and reality. "Slaves sing most," he writes at the end of chapter 2, "when they are most unhappy. . . . The singing of a man cast away upon a desolate island might be as appropriately considered as evidence of contentment and happiness, as the singing of a slave; the songs of the one and of the other are prompted by the same emotion."

Finally, Douglass concludes his second chapter with a discourse on the nature of interpretation, which we could perhaps call the first charting of the black hermeneutical circle and which we could take again as a declaration of the arbitrary relation between a sign and its referent, between the signifier and the signified. The slaves, he writes, "would compose and sing as they went along, consulting neither time nor tune. The thought that came up, came out—if not in the word, [then] in the sound;—and as frequently in the one as in the other." Douglass describes here a certain convergence of perception peculiar only to members of a very specific culture: The thought could very well be embodied nonverbally, in the sound if not in the word. What is more, sound and sense could very well operate at odds to create through tension a dialectical relation. Douglass remarks: "They would sometimes sing the most pathetic sentiment in the most rapturous tone, and the most rapturous sentiment in the most pathetic tone. . . . They would thus sing as a chorus to words which to many would seem unmeaning jargon, but which, nevertheless, were full of meaning to themselves." Yet the decoding of these cryptic messages did not, as some of us have postulated, depend on some sort of mystical union with their texts. "I did not, when a slave," Douglass admits, "understand the deep meaning of those rude and apparently incoherent songs." "Meaning," on the contrary, came only with a certain aesthetic distance and an acceptance of the critical imperative. "I was myself within the circle," he concludes, "so that I neither saw nor heard as those without might see and hear." There exists always the danger, Douglass seems to say, that the meanings of nonlinguistic signs will seem "natural"; one must view them with a certain detachment to see that their meanings are in fact merely the "products" of a certain culture, the result of shared assumptions and conventions. Not only is meaning culture-bound and the referents of all signs an assigned relation, Douglass tells us, but *how* we read determines *what* we read, in the truest sense of the hermeneutical circle.

The Text Was Meant to Be Preached

Robert G. O'Meally

Typically, scholars and teachers dealing with Frederick Douglass's *Narrative of the Life of an American Slave* (1845) are concerned with the crucial issue of religion, because the tensions and ironies generated by the sustained contrast between white and black religions constitute a vital "unity" in the work. Slavery sends Old Master to the devil, while the slave's forthright struggle for freedom is a noble, saving quest. Douglass's search for identity—paralleling the search of many and varied American autobiographers before him—is tightly bound with his quest for freedom and for truth. The *Narrative* presents scholars and teachers with a variety of religious questions. How does Douglass reconcile his professed Christianity with his evidently pagan faith in Sandy Jenkins's root? Why does Christian Douglass condone (even applaud!) the slaves' constant "sinning" against (lying to, stealing from, even the threatened killing of) the upholders of slavery? What is suggested by the fact that the most fervently religious whites treat their slaves more barbarously than do even the "unsaved" whites? While such topics are integral to a discussion of Douglass's *Narrative* and its relation to religion, they leave untouched a vital dimension of this broad subject.

The *Narrative* does more than touch upon questions often pondered by black preachers. Its very form and substance are directly

---

From *Afro-American Literature: The Reconstruction of Instruction,* edited by Dexter Fisher and Robert B. Stepto. © 1978 by the Modern Language Association of America.

influenced by the Afro-American preacher and his vehicle for ritual expression, the sermon. In this sense, Douglass's *Narrative* of 1845 is a sermon, and, specifically, it is a black sermon. This is a text meant to be read and pondered; it is also a Clarion call to spiritual affirmation and action: This is a text meant to be preached.

II

The Afro-American sermon is a folkloric process. More than a body of picturesque items for the catalog, the black sermon is a set of oratorical conventions and techniques used by black preachers in the context of the Sunday morning (or weeknight revival) worship service. The black sermon—especially as delivered in churches of independent denominations, which developed in relative isolation from white control—is distinctive in structure, in diction, and in the values it reflects.

Certain aspects of the black sermon's structure vary greatly from preacher to preacher; indeed, the black congregation expects its preacher to have idiosyncrasies in his manner and form of presentation. In keeping with the thinking of the seemingly remote American Puritans, black church men and women view their preacher's personal style and "voice" as bespeaking his discovery of a personal Christian identity and a home in Christiandom; each telling of The Story is as different in detail as each individual teller. In shaping his sermon, a black preacher may follow the American Puritan formula: doctrine, reasons, uses. Or he may use a historical, an analytical, or a narrative scheme for organizing his presentation of the World. In any case, most Afro-American preachers pace themselves with care, beginning slowly, perhaps with citations from the Bible, or with a prayer, or with a deliberate statement and restatement of the topic for the day.

Most black preachers also build toward at least one ringing crescendo in their sermons, a point when their words are rhythmically sung or chanted in a modified, "ritual" voice. Here the call-response pattern is most marked; the preacher's words are answered by the congregation's phrases, "All right!" "Yes, brother!" "Say *that!*" Sometimes the preacher will rock in rhythm and chant visions of golden heaven and warnings of white-hot hell to his listeners. Sometimes he becomes "laughing-happy" as he walks the pulpit, declaring in words half-sung, half-spoken, how glad he is to be saved by the grace of the Lord. In this crescendo section of the sermon, the highly rhythmical language is closer to poetry than it is to prose. Such chanting may occur only at the conclusion of the sermon, or there may be several such poetical sections. In them the preacher seems possessed; the words are not his own, but the Spirit's.

Classic rhetorical and narrative techniques also abound in the Afro-American sermon. One notes the rich use of metaphors and figures of speech, such as repetitions, apostrophes, puns, rhymes, and hyperboles. A good preacher will not just report as a third-person narrator what the Bible says, but he will address the congregation as a first-person observer: "I can see John," the preacher might say, "walking in Jerusalem *early* one Sunday morning. He is a master of rhetorical and narrative devices.

Characteristically, too, Afro-American sermons are replete with stories from the Bible, folklore, current events, and virtually any source whatsoever. Whether or not this storytelling aspect of the black preacher is an African "survival," as some researchers claim, the consistent use of stories determines the black sermon's characteristic structure. Some stories may provide the text for a sermon, while others occur repeatedly as background material.

James Weldon Johnson notes that certain narrative "folk sermons" are repeated in pulpits Sunday after Sunday. Or, a section of one well-known sermon may be affixed to another sermon. Some of these "folk sermons" include "The Valley of Dry Bones," based on Ezekiel's vision; the "Train Sermon," in which God and Satan are portrayed as train conductors transporting saints and sinners to heaven and to hell; and the "Heavenly March," featuring man on his lengthy trek from a fallen world to a heavenly home. Johnson's own famous poem, "The Creation," is based on another "folk sermon" in which the preacher narrates the story of the world from its birth to the day of final Judgment.

Black sermons are framed in highly figurative language. Using tropes, particularly from the Bible, spirituals, and other sermons, the black preacher's language—especially in chanted sections of his sermon—is often dramatic and full of imagery. In one transcription of a black sermon, a preacher speaks in exalted language of the Creator's mightiness:

> I vision God wringing
> A storm from the heavens

>Rocking the world
>Like an earthquake;
>Blazing the sea
>Wid a trail er fire.
>His eye the lightening's flash,
>His voice the thunder's roll.
>Wid one hand He snatched
>The sun from its socket,
>And the other He clapped across the moon.
>I vision God standing
>On a mountain
>Of burnished gold,
>Blowing his breath
>Of silver clouds
>Over the world,
>His eye the lightening's flash,
>His voice the thunder's roll.

Like other American preachers, the black preacher speaks to his listeners' hearts as well as to their minds. He persuades his congregation not only through linear, logical argumentation but also through the skillful painting of word pictures and the dramatic telling of stories. His tone is exhortative: He implores his listeners to save themselves from the flaming jaws of hell and to win a resting place in heaven. The black preacher may speak in mild, soothing prose, or he may, filled with the spirit, speak in the fiery, poetical tongue of the Holy Spirit. The black preacher's strongest weapon against the devil has been his inspired use of the highly conventionalized craft of sacred black oratory—a folkloric process.

III

The influences of the black sermon on black literature have been direct and constant. The Afro-American playwright, poet, fiction writer, and essayist have all drawn from the Afro-American sermon. Scenes in black literature occur in church; characters recollect particularly inspiring or oppressive sermons; a character is called upon to speak and falls into the cadences of the black sermon, using the familiar Old Testament black sermonic stories and images. In his essays, James Baldwin, who preached when he was in his teens, employs the techniques of the sermon as he speaks to his readers' hearts and souls about their sins and their hope for salvation. Just as one finds continuity in tone and purpose from the sermons of the Puritans to the essays of such writers as Emerson and Thoreau, one discovers continuity in the Afro-American literary tradition from the black sermon—still very much alive in the black community—to the Afro-American narrative, essay, novel, story, and poem.

What, then, is *sermonic* about Douglass's *Narrative?* First of all, the introductory notes by William Lloyd Garrison and Wendell Phillips, both fiery orators and spearheads of the abolition movement, prepare the reader for a spiritual message. In his preface, Garrison recalls Douglass's first speech at an antislavery convention. Thunderous applause follows the ex-slave's words, and Garrison says, "I never hated slavery so intensely as at that moment; certainly my perception of the enormous outrage which is inflicted by it, on the godlike nature of its victims was rendered far more clear." And then, in stormy, revivalist style, Garrison rises and appeals to the convention, "whether they would ever allow him [Douglass] to be carried back into slavery,—law or no law, constitution or no constitution. The response was unanimous and in thunder—tones—'No!' 'Will you succor and protect him as a brother-man—a resident of the old Bay State?' 'Yes!' shouted the whole mass."

As if introducing the preacher of the hour, Garrison says that Douglass "excels in pathos, wit, comparison, imitation, strength of reasoning, and fluency of language." Moreover, in Douglass one finds "that union of head and heart, which is indispensable to an enlightenment of the heads and winning the hearts of others. . . . May he continue to 'grow in grace, and in the knowledge of God' that he may be increasingly serviceable in the cause of bleeding humanity, whether at home or abroad." As for Douglass's present narrative, says Garrison, it grips its readers' hearts:

> He who can peruse it without a tearful eye, a heaving breast, an afflicted spirit,—without being filled with an unutterable abhorrence of slavery and all its abettors, and animated with a determination to seek the immediate overthrow of that execrable system,—without trembling for the fate of this country in the hands of a righteous God, who is ever on the side of the oppressed, and whose arm is not shortened that it cannot save,—must have a flinty

heart, and be qualified to act the part of a trafficker "in slaves and the souls of men."

The choices, Garrison states, are but two: enrollment in the righteous war against slavery or participation in the infernal traffic in "the souls of men."

In his turn, Wendell Phillips prepares the way for Douglass's "sermon." In his laudatory letter to the author, Phillips speaks of Southern white slave masters as infrequent "converts." Most often, the true freedom fighter detests slavery in his heart even "before he is ready to lay the first stone of his anti-slavery life." Phillips thanks Douglass especially for his testimony about slavery in parts of the country where slaves are supposedly treated most humanely. If things are so abominable in Maryland, says Phillips, think of slave life in "that Valley of the Shadow of Death, where the Mississippi sweeps along."

Douglass's account of his life serves the ritual purpose announced in the prefatory notes: The ex-slave comes before his readers to try to save their souls. His purpose is conversion. In incident upon incident, he shows the slaveholder's vile corruption, his lust and cruelty, his appetite for unchecked power, his vulgarity and drunkenness, his cowardice, and his damning hypocrisy. Slavery, says Douglass, brings sin and death to the slaveholder. Come to the abolition movement, then, and be redeemed. Take, as Douglass has done, the abolitionist paper as a Bible and freedom for all men as your heaven. Addressed to whites, the *Narrative* is a sermon pitting the dismal hell of slavery against the bright heaven of freedom.

Douglass's portrayal of himself and of his fellow slaves is in keeping with the text's ritual function. Like a preacher, he has been touched by God, *called* for a special, holy purpose. Providence protects Douglass from ignorance and despair. Providence selects him to extend his vision of freedom and, concretely, to move to Baltimore. The unexplained selection of Douglass to go to Baltimore he sees as "the first plain manifestation of that kind providence which has ever attended me, and marked my life with so many favors." Of this "providential" removal to Baltimore Douglass further writes:

> I may be deemed superstitious, and even egotistical, in regarding this event as a special interposition of divine Providence in my favor. But I should be false to the earliest sentiments of my soul, if I suppressed the opinion. . . .

> From my earliest recollection, I date the entertainment of a deep conviction that slavery would not always be able to hold me within its foul embrace; and in the darkest hours of my career in slavery, this living word of faith and spirit of hope departed not from me, but remained like ministering angels to cheer me through the gloom. This good spirit was from God, and to him I offer thanksgiving and praise.

In his effort to convert white slaveholders and to reassure white abolitionists, Douglass attempts to refute certain racist conceptions about blacks. He presents blacks as a heroic people suffering under the lash of slavery but struggling to stay alive to obtain freedom. To convince whites to aid slaves in their quest for freedom Douglass tackles the crude, prejudiced assumptions—which slavers say are upheld by Scripture—that blacks somehow *deserve* slavery, that they enjoy and feel protected under slavery. Of the notion that blacks are the cursed descendants of Ham, Douglass writes, "if the lineal descendants of Ham are alone to be scripturally enslaved, it is certain that slavery at the south must soon become unscriptural; for thousands are ushered into the world, annually, who, like myself, owe their existence to white fathers, and those fathers most frequently their own masters." Furthermore, if cursed, what of the unshakable conviction of the learned and eloquent Douglass that he is, in fact, chosen by God to help set black people free?

What, then, of the assumption of the plantation novel and the minstrel show that blacks are contented with "their place" as slaves at the crushing bottom of the American social order? Douglass explains that a slave answers affirmatively to a stranger's question, "Do you have a kind master?" because the questioner may be a spy hired by the master. Or the slave on a very large plantation who complains about his master to a white stranger may later learn that the white stranger was, in fact, his master. One slave makes this error with Colonel Lloyd, and, in a few weeks, the complainer is told by his overseer that, for finding fault with his master, he is now being sold into Georgia. Thus, if a slave says his master is kind, it is because he has learned the maxim among his brethren "A still tongue makes a wise head." By suppressing the truth rather than taking the consequences of telling it foolishly, slaves "prove themselves a part of the human family."

At times, slaves from different plantations may argue or even fight over who has the best, the kindest, or the manliest master. "Slaves are like other people and imbibe prejudices quite common to others," explains Douglass. "They think their own better than that of others." Simultaneously, however, slaves who publicly uphold their masters' fairness and goodness, "execrate their masters" privately.

Do not the slaves' songs prove their contentedness and joy in bondage? "It is impossible," says Douglass, "to conceive of a greater mistake." Indeed, he says,

> The songs of the slave represent the sorrows of his heart; and he is relieved by them, only as an aching heart is relieved by its tears. At least, such is my experience. I have often sung to drown my sorrow, but seldom to express my happiness. Crying for joy, and singing for joy, were alike uncommon to me while in the jaws of slavery.

Instead of expressing mirth, these songs Douglass heard as a slave "told a tale of woe which was then altogether beyond my feeble comprehension; they were tones loud, long, and deep; they breathed the prayer and complaint of souls boiling over with the bitterest anguish. Every tone was a testimony against slavery, and a prayer to God for deliverance from chains." These songs, Douglass recalls, gave him his "first glimmering conception of the dehumanizing character of slavery." In other words, these songs "prove" the black man's deep, complex humanity. Therefore, whites, come forth, implies Douglass, and join the fight to free these God's children!

(The tone of Douglass's *Narrative* is unrelentingly exhortative. Slaveholders are warned that they tread the road toward hell, for even as their crimes subject the slave to misery, they doom the master to destruction. Douglass describes his aged grandmother's abandonment in an isolated cabin in the woods. Then in dramatic, rhythmical language, he warns that,

> My poor old grandmother, the devoted mother of twelve children, is left all alone, in yonder hut, before a few dim embers. She stands—she sits—she staggers—she falls— she groans—she dies—and there are none of her children or grandchildren present, to wipe from her wrinkled brow the cold sweat of death, or to place beneath the sod her

fallen remains. Will not a righteous God visit for these things?

Later in the text, Douglass exhorts white readers to sympathize with the escaping slave's plight. To comprehend the escapee's situation, the white sympathizer "must needs experience it, or imagine himself in similar circumstances." In a voice one imagines to be as strong and varied in pitch as a trombone, Douglass reaches a crescendo, in black sermon style, when speaking in highly imagistic language of the white man who would comprehend the escaped slave's feelings:

> Let him be a fugitive slave in a strange land—a land given up to be the hunting ground for slaveholders—whose inhabitants are legalized kidnappers—where he is every moment subjected to the terrible liability of being seized upon by his fellowmen, as the hideous crocodile seizes upon his prey!—I say, let him place himself in my situation—without home or friends—without money or credit—wanting shelter, and no money to buy it,—and at the same time let him feel that he is pursued by merciless men-hunters, and in total darkness as to what to do, where to go, or where to stay,—perfectly helpless both as to the means of defence and means of escape,—in the midst of plenty, yet suffering the terrible gnawings of hunger,—in the midst of houses, yet having no home,—among fellowmen, yet feeling as if in the midst of wild beasts, whose greediness to swallow up the trembling and half-famished fugitive is only equalled by that with which the monsters of the deep swallow up the helpless fish upon which they subsist,—I say, let him be placed in that most trying situation,—the situation in which I was placed,—then, and not till then, will he fully appreciate the hardships of, and know how to sympathize with, the toil-worn and whip-scarred fugitive slave.

In this passage Douglass, like a black preacher, uses a variety of oratorical techniques: alliteration, repetition, parallelism. Also, using conjunctions, commas, and dashes, Douglass indicates the dramatic pauses between phrases and the surging rhythms in the sermonlike prose.

Like a sermon, too, Douglass's *Narrative* argues not only by stern reason but also with tales that may be termed *parables*. One of the most forceful of these parables, one threaded quite successfully into the *Narrative,* is the parable of poor Mrs. Auld. Residing in the border state of Maryland, in the relatively large city of Baltimore, Mrs. Auld, who has never owned a slave before she owns Frederick Douglass, is truly a good woman. Before her marriage, Mrs. Auld worked as a weaver, "dependent upon her own industry for a living." When eight-year-old Douglass is brought into the Auld household, Mrs. Auld is disposed to treat him with human respect and kindness. Indeed, "her face was made of heavenly smiles, and her voice was tranquil music." Douglass obviously presents this woman as a glowing model of Christian charity: "When I went there," he writes, "she was a pious, warm, and tender-hearted woman. There was no sorrow or suffering for which she had not a tear. She had bread for the hungry, clothes for the naked, and comfort for every mourner within her reach." Soon after Douglass arrives in her home, Mrs. Auld begins to do as she has done for her own son; she commences teaching Douglass the alphabet.

Before long, of course, this "kind heart" is blasted by the "fatal poison of irresponsible power." In Douglass's words, Mrs. Auld's "cheerful eye, under the influence of slavery, soon became red with rage; that voice, made all of sweet accord, changed to one of harsh and horrid discord; and that angelic face gave place to that of a demon." In response to her husband's warning that education "would *spoil* the best nigger in the world," she forbids Douglass's further instruction. In fact, she becomes at last "even more violent than her husband himself" in the application of this precept that slave education is a danger. Thus, even the mildest forms of slavery—in providential Baltimore—turn the most angelic face to that of a "harsh and horrid" devil.

The central paradox of the story of Mrs. Auld is that Mr. Auld's vitriolic warning against learning actually serves to make Douglass double his efforts to gain literacy. Mr. Auld's words to his wife prove prophetic:

> "Now," said he, "if you teach that nigger . . . how to read, there would be no keeping him. It would forever unfit him to be a slave. He would at once become unmanageable, and of no value to his master. As to himself, it

could do him no good, but a great deal of harm. It would make him discontented and unhappy.

Douglass overhears this warning and feels that at last he comprehends the source of the white man's power to enslave blacks. "From that moment," writes Douglass, "I understood the pathway from slavery to freedom." Also, from that moment on, Douglass's holy search for identity and freedom is knotted to his determined quest for literacy and knowledge. For the skill of literacy, "I owe almost as much to the bitter opposition of my master," writes Douglass, "as to the kindly aid of my mistress. I acknowledge the benefit of both." It is as if the providentially guided Douglass receives truth from the mouths of family and friends, and even from the mouths of his most indefatigable enemies. And like a preacher he reports his successes (the Good Word) in exalted prose and in parables.

A second major parable in the *Narrative* concerns the slave-breaker Edward Covey and the wise old slave Sandy Jenkins. Like Mrs. Auld, Covey is a hard worker whose diligence fails to shield him from the blight of slavery. Sent to Covey's plantation to be "broken," Douglass, in a sense, breaks Covey. Douglass leaves the slave-breaker's plantation stronger than ever in his personal resolution to break free. At first, the deceptive Covey, with his killing work schedule and "tiger-like" ferocity, seems to have succeeded in "taming" Douglass. "I was broken," says Douglass, "in body, soul, and spirit. My natural elasticity was crushed, my intellect languished, the disposition to read departed, the cheerful spark that lingered about my eye died; the dark night of slavery closed in upon me; and behold a man transformed into a brute!" On Sundays, his only free day, Douglass would lounge in a "beast-like stupor" under a tree. His thoughts of killing himself and Covey are checked only by fear and dim hope.

Somehow, though, Douglass's spirits are rekindled. First, he observes the white sails of the ships piloting the Chesapeake Bay—through identification with their bold freedom, and through soliloquies to them and to God, Douglass finds his hopes revived. He too will try to sail to freedom. Quoting a line from a spiritual, he says, "There is a better day coming."

Second, he faces down Covey. "You have seen how a man was made a slave," writes Douglass. "You shall see how a slave was made a man." One day he faints and is unable to do his work. Covey orders him to arise and return to his labor, but Douglass says, "I made no effort to comply, having now made up my mind to let him do his worst." Then Douglass runs off to his master, Mr. Thomas Auld, to express fear that Covey will kill him. But Auld merely sends Douglass back to the slave-breaker. Back at Covey's, Douglass is chased into the woods by the slave-breaker, who wields a cowskin. Ordinarily, running away could only make things worse for Douglass: "My behavior," he says, "was altogether unaccountable." Providence seems to be with him though. In the woods, Douglass meets his old acquaintance Sandy Jenkins, who advises him to return to Covey. But Jenkins does not send Douglass back to Covey unarmed. Jenkins directs Douglass to a part of the woods where he can find a certain root, which, Douglass says, "if I would take some of it with me, carrying it *always on my right side,* would render it impossible for Mr. Covey, or any other white man, to whip me." In the years that he has carried his root, Jenkins says, he had never been beaten, and he never expects to be again. Douglass seeks relief from Covey by petition to Auld, then by attempted escape, and then by the spiritual guidance—dependent upon personal fortitude and faith—symbolized by the old slave's root.

Upon his return to Covey's, Douglass is spared an initial attack, presumably because Covey, a leader in his church, does not want to work or whip slaves on Sunday. Monday morning, however, Covey comes forth with a rope and—"from whence came the spirit I don't know," says Douglass—the slave resolves to fight. They fight for nearly two hours, Douglass emerging unscarred and Covey bloodied. This fight marks an important rite of passage for Douglass:

> This battle with Mr. Covey was the turning-point in my career as a slave. It rekindled the few expiring embers of freedom, and revived within me a sense of my own manhood. It recalled the departed self-confidence, and inspired me again with a determination to be free.... I now resolved that, however long I might remain a slave in form, the day had passed forever when I could be a slave in fact. I did not hesitate to let it be known of me, that the white man who expected to succeed in whipping, must also succeed in killing me.
>
> From this time I was never again what might be called fairly whipped, though I remained a slave four years afterwards. I had several fights, but was never whipped.

Thus we see Douglass, Providence's hero, maneuvering through deadly dangers. His straightforwardness and courage defeat the serpentine and Pharisee-like Covey. Douglass's hope returns through identification with the white sails on the Bay. He is also given heart by the root, a symbol of spiritual and natural power as well as of the supreme power of hope and faith.

As in a successful black sermon, these parables are well woven into the whole cloth of the *Narrative*. They illustrate the corrupting power of slavery upon whites; they illustrate the power of the slave to overcome the slaveowner and to return, mysteriously—and by the power of Providence—to the winding road to freedom. Douglass's *Narrative* is alive with allusions to the Bible. Inevitably, the war waged is between the devil of slavery and the righteous, angry God of freedom. Chapter 3 commences with a description of Colonel Lloyd's garden. In its beauty and power to tempt, this "large and finely cultivated garden" recalls the Garden of Eden:

> This garden was probably the greatest attraction of the place. During the summer months, people came from far and near—from Baltimore, Easton, and Annapolis—to see it. It abounded in fruits of almost every description, from the hardy apple of the north to the delicate orange of the south. This garden was not the least source of trouble on the plantation. Its excellent fruit was quite a temptation to the hungry swarms of boys, as well as the older slaves, belonging to the colonel, few of whom had the virtue or the vice to resist it.

This Eden, though carefully tended as God commanded, is vile and corrupt. Colonel Lloyd, merciless owner of the garden and gardeners, forbids the slaves to eat any of its excellent fruits. To enforce his rule he has tarred the garden fence; any slave with tar on his person was deemed guilty of fruit theft and was "severely whipped." This is an Eden controlled not by God but by greedy, selfish, slaveholding man.

Or is this garden under the charge of the devil? As noted, slavery turns the heart of "heavenly" Mrs. Auld to flinty stone. And Mr. Plummer, Douglass's first overseer, is "a miserable drunkard, a profane swearer" known to "cut and slash women's heads so horribly" that even the master becomes enraged. This enraged master, Captain Anthony, *himself* seems "to take great pleasure in whipping a slave." In a grueling scene, he whips Douglass's aunt Hester, a favorite of

Anthony's, until only the master's fatigue stops the gory spectacle. "The louder she screamed, the harder he whipped; and where the blood ran fastest, there he whipped longest." Mr. Severe would curse and groan as he whipped the slave women, seeming "to take pleasure in manifesting this fiendish barbarity." Colonel Lloyd renders especially vicious beatings to slaves assigned to the care of his horses. When a horse "did not move fast enough or hold high enough," the slaves were punished. "I have seen Colonel Lloyd make old Barney, a man between fifty and sixty years of age, uncover his bald head, kneel down upon the cold, damp ground, and receive upon his naked and toil-worn shoulders more than thirty lashes at the time." Other slaveowners and overseers, both men and women, kill their slaves in cold blood.

One of the men termed "a good overseer" by the slaves is Mr. Hopkins, who, at least is not quite so profane, noisy, or cruel as his colleagues. "He whipped, but seemed to take no pleasure in it." His tenure as overseer is a short one, conjectures Douglass, because he lacks the brutality and severity demanded by the master.

Covey is the most devil-like slaveholder in the *Narrative*. Hypocritical, masterful at deception, clever, untiring, seemingly omnipresent, Covey is called "The Snake" by the slaves. In one of the *Narrative*'s most unforgettable portraits, Douglass tells us that through a cornfield where Covey's slaves work, The Snake would

> sometimes crawl on his hands and knees to avoid detection, and all at once he would rise nearly in our midst, and scream out, "Ha, ha! Come, come! Dash on, dash on!" This being his mode of attack, it was never safe to stop a single minute. His comings were like a thief in the night. He appeared to us as being ever at hand. He was under every tree, behind every stump, in every bush, and at every window, on the plantation. He would sometimes mount his horse, as if bound to St. Michael's, a distance of seven miles, and in half an hour afterwards you would see him coiled up in the corner of the woodfence, watching every motion of the slaves.

The Snake has built his reputation on being able to reduce spirited men like Douglass to the level of docile, manageable slaves. Under Covey's dominion—before Douglass gains a kind of dominion over Covey—Douglass feels himself "transformed into a brute." In the symbolic geography of this text the Garden is ruled—at least for the moment—by none other than his majesty, the infernal Snake, Satan.

Douglass makes clear that slavery, not the slaveowner, is the supreme Devil in this text: slavery, with its "robes crimsoned with the blood of millions." Mrs. Auld falls from "heavenliness" to the hell of slavery. As a boy Douglass learns from Sheridan's speeches in behalf of Catholic emancipation that "the power of truth [holds sway] over the conscience of even a slaveholder." These white slaveholders, if devil-like, are nonetheless capable of redemption.

Colonel Lloyd, in fact, is described as possessing wealth equal almost to that of Job—the Old Testament's model of supreme faithfulness. Finally, however, the effect of the comparison is ironical, for Lloyd is a man of increasing cruelty; his very wealth seems to provide his temptation to do evil, and Lloyd yields to temptation with relish.

There are several places in the *Narrative* where American slavery is compared with the holding of the Old Testament Jews in captivity. Douglass points out that, the more he read, the more he viewed his enslavers as "successful robbers, who had left their homes, and gone to Africa, and stolen us from our homes, and *in a strange land* reduced us to slavery" (italics mine). Later in the *Narrative* Douglass describes the fugitive slave in the North as a dweller "in a strange land." The language here and the parallel situations recall the biblical psalm that reads

> By the rivers of Babylon, there we
> sat down, yea, we wept, when
> we remembered Zion
>   we hanged our harps upon the
> willows in the midst thereof.
>   For there they that carried us away
> captive required of us a song; and
> they that wasted us required of us
> mirth, saying, Sing us one of the
> songs of Zion
>   How shall we sing the Lord's
> Song in a strange land?

This allusion is even more suggestive when one considers Douglass's careful explanation of the slaves' songs, demanded, in a sense, and misunderstood by the captors, who, thinking the songs joyous, feel the more justified in their ownership of the black singers.

[In the *Narrative,* Douglass calls on the Old Testament God to free His black children.] "For what does he hold the thunders in his right hand, if not to smite the oppressor, and deliver the spoiled out of the hand of the spoiler?" Douglass also manifests certain characteristics of an Old Testament hero. He becomes Daniel, blessed with supernatural powers of perception and protected by God's special favor. Like Daniel, thrown into a den of lions for refusing to refrain from praying, Douglass never loses faith while he lives in the very "jaws of slavery." Upon being returned from the Lloyd Plantation to Baltimore, Douglass felt he had "escaped a worse than lion's jaws." Captain Auld in St. Michael's was a vicious, ineffectual master who, Douglass tells us, "Might have passed for a lion, but for his ears." Escaping for a brief time from Covey, Douglass, sick and scarred, returns to Auld. The runaway slave supposes himself to have "looked like a man who had escaped a den of wild beasts, and barely escaped them." Unlike Daniel, Douglass actually has to battle with the lions, tigers, and "the Snake" in the den of slavery. Like Daniel, though, he is protected, and once he has Sandy's root on his right side he can never be beaten. Providentially, too, Douglass is eventually rescued from the crushing jaws of slavery.

(Douglass's account of his life follows the pattern of the life of a mythic or historic hero—or a hero of Scripture. His birth, if not virginal, as is so often the case with the archetypal hero, is cloaked in mystery.) He is never sure who his father is or even when, exactly, he himself was born. Nor does he feel very close to his natural family; slavery kept mother from son, and brother from sister, so that natural familial bonds were felt only remotely. Like Joseph (the biblical son sold into slavery by his brothers) and like Moses, Douglass feels sure he has been selected by heaven for special favor. And like Jesus, he prays for redemption and resurrection from "the coffin of slavery to the heaven of freedom." Christ-like, too, is Douglass's faltering faith at the torturous nadir of his enslavement. Under Covey's lash, Douglass nearly surrenders to the bestial slave system, and to murder and suicide. But Douglass turns from the false religion of such "Pharisees" as Covey; like Jesus, Douglass criticizes white institutionalized worship but clings to his faith in a personal Father.

Douglass's personal sense of ethics contradicts the codes of such men as Covey. For instance, Douglass hails the slave's trickery of the master as wit, if not wisdom. Douglass also approaches the attempted assassination of a black informer on runaway slaves; such is justice. Moreover, although Douglass disclaims "ignorant" and "superstitious" belief in the power of the root, his true feeling about root power emerges from the *Narrative*. Clearly, the root, be it pagan or nonpagan, gives Douglass the strength to master Covey. This "superstition" seems no contradiction in Douglass, for he is presented as a hero who transcends strict adherence to existing law. He is the possessor of pure religion; God speaks directly to him. Like a Christ or a Moses, he not only follows God's law, he *gives* the law. Clearly, this pure, felt religion of real experience with Providence is not the religion of the white slaveholding churchmen who merely use Christianity to justify their crimes.

Douglass's rejection of the slaveholder's false religion parallels the rejection of popular conceptions of God by such diverse American writers as Benjamin Franklin, Thomas Paine, and—writing a full century after the *Narrative*'s publication—James Baldwin. Like these writers, Douglass replaces the hollow religion of form for a deep, personal religion—in his case, the religion of abolition, which he practices and preaches with fervent passions.

Furthermore, like many black preachers, Douglass's true religion is a practical one that seeks a "heaven" on earth as well as on high. Salvation is not only a personal matter; Douglass labors for the freedom of a *people*. Once free (or at least freer) in Massachusetts, he joins the abolition movement: "It was a severe cross," he writes, "and I took it up reluctantly." His *Narrative* is, then, not only the spiritual journey of one soul but also a testimony and a warning, written with the earnest hope that it "may do something toward throwing light on the American slave system, and hastening the glad day of deliverance to the millions of my brethren." Like a black sermon, it is the story of a people under the guidance of Providence.

Douglass's message is the message of the progressive black preacher: Be hopeful and faithful, but do not fail to fight for the freedom of your brother men. Douglass recognizes that the God of freedom respects the slave who may lie, cheat, steal, or even kill to stay alive and to struggle for freedom. This freedom ethic, "preached" by Douglass, was in the tradition of many militant black preachers, including the black preacher and pamphleteer, Rev. Henry Highland Garnet.

Douglass's *Narrative* is in its way, a holy book—one full of marvels, demonstrating God's active participation in a vile and fallen world. The *Narrative* is a warning of the terror of God's fury. It is also an account of a black Moses' flight "from slavery to freedom." It is an invitation to join "the church" of abolition, a church that offers freedom not only to the slave and the sympathetic white Northerner but also to the most murderous and bloodthirsty Southern dealers in human flesh. Sinners, Douglass seems to chant, black sermon-style, you are in the hands of an angry God!

Clearly, this is an autobiography, a slave narrative, a fictionlike work shaped by oratory as well as the sentimental romance. But Douglass, who grew up hearing sermons on the plantation and who heard and delivered them throughout his life, produced, in this greatest account of his life, a text shaped by the form and the processes of speaking characteristic of the black sermon. This is a mighty text meant, of course, to be read. But it is also a text meant to be mightily preached.

Autobiographical Acts and the Voice of the Southern Slave

Houston A. Baker, Jr.

The Southern slave's struggle for terms for order is recorded by the single, existential voice engaged in what Elizabeth Bruss calls "autobiographical acts." How reliable are such acts? Benedetto Croce called autobiography "a by-product of an egotism and a self-consciousness which achieve nothing but to render obvious their own futility and should be left to die of it." And a recent scholar of black autobiography expresses essentially the same reservations: "Admittedly, the autobiography has limitations as a vehicle of truth. Although so long an accepted technique towards understanding, the self-portrait often tends to be formal and posed, idealized or purposely exaggerated. The author is bound by his organized self. Even if he wishes, he is unable to remember the whole story or to interpret the complete experience." A number of eighteenth- and nineteenth-century American thinkers would have taken issue with these observations. Egotism, self-consciousness, and a deep and abiding concern with the individual are at the forefront of American intellectual traditions, and the formal limitations of autobiography were not of great concern to those white authors who felt all existent literary forms were inadequate for representing their unique experiences. The question of the autobiography's adequacy, therefore, entails questions directed not only toward the black voice in the South, but also toward the larger context of the American experiment as a whole.

From *The Journey Back: Issues in Black Literature and Criticism.* © 1975 by Houston A. Baker, Jr. University of Chicago Press, 1980.

Envisioning themselves as God's elect and imbued with a sense of purpose, the Puritans braved the Atlantic on a mission into the wilderness. The emptiness of the New World, the absence of established institutions and traditions, reinforced their inclination to follow the example of their European forebears and brothers in God. They turned inward for reassurance and guidance. Self-examination became the sine qua non in a world where some were predestined for temporal leadership and eventual heavenly reward and others for a wretched earthly existence followed by the fires of hell. The diary, the journal, the meditation, the book of evidences drawn from personal experiences were the literary results of this preoccupation with self, and even documents motivated by religious controversy often took the form of apology or self-justification. A statement from Jonathan Edwards's *Personal Narrative* offers a view of this tradition: "I spent most of my time in thinking of divine things, year after year: often walking alone in the woods, and solitary places, for meditation, soliloquy, and prayer, and converse with God; and it was always my manner at such time, to sing forth my contemplations."

The man alone, seeking self-definition and salvation, certain that he has a God-given duty to perform, is one image of the white American writer. Commenting on Edwards and the inevitable growth of autobiography in a land without a fully articulated social framework, Robert Sayre writes: "Edwards could and had to seek self-discovery within himself because there were so few avenues to it outside himself. The loneliness and the need for new forms really go together. They are consequences of one another and serve jointly as inducements and as difficulties to autobiography." This judgment must be qualified, since Edwards's form does not differ substantially from John Bunyan's, and his isolated meditations fit neatly into a Calvinistic spectrum, but Sayre is fundamentally correct when he specifies a concern with solitude and a desire for unique literary expression as key facets of the larger American experience.

Despite the impression of loneliness left by Edwards and the sense of a barren and unpromising land for literature left by comments like those of Hawthorne in his preface to *The Marble Faun* or James in *Hawthorne,* there were a number of a priori assumptions available to the white American thinker. They developed over a wide chronological span (the original religious ideals becoming, like those treated in the discussion of black writers above, increasingly secular) and provided a background ready to hand. There was the white writer's sense that he was part of a new cultural experience, that he had gotten away from what D. H. Lawrence calls his old masters and could establish a new and fruitful way of life in America. There was the whole panoply of spiritual sanctions; as one of the chosen people, he was responsible for the construction of a new earthly paradise, one that would serve as a holy paradigm for the rest of the world. There was the white writer's belief, growing out of the liberal, secular thought of Descartes, Locke, and Newton, that the individual was unequivocally responsible for his own actions; a man was endowed with inalienable rights, and one of these was the right to educate himself and strive for commercial success. There was also the feeling that America offered boundless opportunities for creative originality: a unique culture with peculiar sanctions should produce a sui generis art.

Thus, while James's "extraordinary blankness—a curious paleness of colour and paucity of detail" was characteristic for some early white Americans, there were also more substantial aspects or qualities of the American experience that stood in contrast to this "blankness." The writer could look to a Puritan ontology and sense of mission, to conceptions of the self-made man, or to a prevailing American concern for unique aesthetic texts as preshaping influences for his work. The objective world provided both philosophical and ideological justifications for his task. When Emerson wrote, "Dante's praise is that he dared to write his autobiography in colossal cipher, or into universality," he optimistically stated the possibilities immanent in the white author's situation. The writer of comprehensive soul who dared to project his experiences on a broad plane would stand at the head of a great tradition. According to Emerson, the world surrounding such a person—that supposedly void externality—offered all the necessary supports. The permanence and importance of works such as Edwards's *Personal Narrative,* Whitman's *Leaves of Grass,* and Adams's *The Education of Henry Adams* in American literature confirm his insight. As the American autobiographer turned inward to seek "the deepest *whole* self of man" (Lawrence's phrase), he carried with him the preexistent codes of his culture. They aided his definition of self and are fully reflected in the resultant texts—self-conscious literary autobiographies.

This perspective on white American autobiography highlights the distinctions between two cultures. Moved to introspection by the apparent "blankness" that surrounded him, the black, Southern field slave had scarcely any a priori assumptions to act as stays in his quest for self-definition. He was a man of the diaspora, a displaced person imprisoned by an inhumane system. He was among alien gods in a strange land. Vassa describes his initial placement in the New World:

> We were landed up a river a good way from the sea, *about Virginia country,* where we saw few or none of our native Africans, and not one soul who could talk to me. I was a few weeks weeding grass and gathering stones in a plantation; and at last all my companions were distributed different ways, and only myself was left. I was now exceedingly miserable, and thought myself worse off than any of the rest of my companions, for they could talk to each other, but I had no person to speak to that I could understand. In this state, I was constantly grieving and pining, and wishing for death rather than anything else.

For the black slave, the white externality provided no ontological or ideological certainties; in fact, it explicitly denied slaves the grounds of being. The seventeenth- and eighteenth-century black codes defined blacks as slaves in perpetuity, removing their chance to become free, participating citizens in the American city of God. The Constitution reaffirmed the slave's bondage, and the repressive legislation of the nineteenth century categorized him as "chattel personal." Instead of the ebullient sense of a new land offering limitless opportunities, the slave, staring into the heart of whiteness around him, must have felt as though he had been flung into existence without a human purpose. The white externality must have loomed like the Heideggerian "nothingness," the negative foundation of being. Jean Wahl's characterization of Heidegger's theory of existence captures the point of view a black American slave might justifiably have held: "Man is in this world, a world limited by death and experienced in anguish; is aware of himself as essentially anxious; is burdened by his solitude within the horizon of his temporality."

There were at least two alternatives to this vision. There was the recourse of gazing idealistically back to "Guinea." Sterling Stuckey has shown that a small, but vocal, minority of blacks have always employed this strategy. And we have already considered its employment in the work of a Northern spokeswoman like Wheatley or a black abolitionist like Vassa. There was also the possibility of adopting the God of the enslaver as solace. A larger number of blacks chose this option and looked to the apocalyptic day that would bring their release from captivity and vengeance on the oppressors. (Tony McNeill's words, "between Africa and heaven," come to mind.) Finally, though, the picture that emerges from the innumerable accounts of slaves is charged with anguish—an anguish that reveals the black bondsman to himself as cast into the world, forlorn and without refuge.

And unlike white Americans who could assume literacy and familiarity with existing literary models as norms, the slave found himself without a system of written language—"uneducated," in the denotative sense of the word. His task was not simply one of moving toward the requisite largeness of soul and faith in the value of his experiences. He first had to seize the word. His being had to erupt from nothingness. Only by grasping the word could he engage in the speech acts that would ultimately define his selfhood. Further, the slave's task was primarily one of creating a human and liberated self rather than of projecting one that reflected a peculiar landscape and tradition. His problem was not to answer Crèvecoeur's question: "What then is the American, this new man?" It was, rather, the problem of being itself.

The *Narrative of the Life of Frederick Douglass,* one of the finest black American slave narratives, serves to illustrate the black autobiographer's quest for being. The recovered past, the journey back, represented in the work is a sparse existence characterized by brutality and uncertainty:

> I have no accurate knowledge of my age. The opinion was . . . whispered about that my master was my father; but of the correctness of this opinion, I know nothing.
>
> My mother and I were separated when I was but an infant.
>
> I was seldom whipped by my old master, and suffered little from anything else than hunger and cold.
>
> Our food was coarse corn meal boiled. This was called *mush.* It was put into a large wooden trough, and set down upon the ground. The children were then called, like so many pigs, and like so many pigs they would come out and devour the mush.

Unlike David Walker who, in his *Appeal,* attempts to explain why blacks are violently held in bondage, the young Douglass finds no explanation for his condition. And though he does describe the treatment of fellow slaves (including members of his own family), the impression left by the first half of the *Narrative* is one of a lone existence plagued by anxiety. The white world rigorously suppresses all knowledge and action that might lead the narrator to a sense of his humanity.

The total process through which this subjugation is achieved can be seen as an instance of the imposed silence suggested by Forten's address. Mr. Hugh Auld, whom Douglass is sent to serve in Baltimore, finding that his wife—out of an impulse to kindness rare among whites in the *Narrative*—has begun to instruct the slave in the fundamentals of language, vociferously objects that "learning would *spoil* the best nigger in the world." Not only is it illegal to teach slaves, but it is also folly. It makes them aspire to exalted positions. The narrator's reaction to this injunction might be equated with the "dizziness" that, according to Heidegger, accompanies a sudden awareness of possibilities that lie beyond anguish:

> These words sank into my heart, stirred up sentiments within that lay slumbering, and called into existence an entirely new train of thought. It was a new and special revelation, explaining dark and mysterious things, with which my youthful understanding had struggled, but struggled in vain. I now understood what had been to me a most perplexing difficulty—to wit, the white man's power to enslave the black man.

Douglass had come to understand, by the "merest accident," the power of the word. His future is determined by this moment of revelation: he resolves, "at whatever cost of trouble, to learn how to read." He begins to detach himself from the white externality around him, declaring:

> What he [Mr. Auld] most dreaded, that I most desired. What he most loved, that I most hated. That which to him was a great evil, to be carefully shunned, was to me a great good to be diligently sought; and the argument which he so warmly urged, against my learning to read, only served to inspire me with a desire and determination to learn.

The balanced antithesis of the passage is but another example—an explicit and forceful one—of the semantic competition involved in culture contact. Mr. Auld is a representation of those whites who felt that by superimposing the cultural sign *nigger* on vibrant human beings like Douglass, they would be able to control the meanings and possibilities of life in America. One marker for the term *nigger* in Auld's semantic field is <<subhuman agency of labor>>. What terrifies and angers the master, however, is that Douglass's capacities—as revealed by his response to Mrs. Auld's kindness and instructions—are not accurately defined by this marker. For Douglass and others of his group are capable of learning. Hence, the markers in Auld's mapping of *nigger* must also include <<agent capable of education>>. The semantic complexity, indeed the wrenching irony, of Auld's "nigger" is forcefully illustrated by the fact that the representation of Auld and *his* point of view enters the world of the learned by way of a narrative written by a "nigger." Douglass, that is to say, ultimately controls the competition among the various markers of *nigger* because he has employed meanings (e.g., agent having the power of literacy) drawn from his own field of experience to represent the competition in a way that invalidates <<subhuman agency of labor>>. The nature of the autobiographical act, in this instance, is one of self-enfolding ironies. Douglass, the literate narrator, represents a Douglass who is perceived by Auld as a "nigger." Certainly the narrator himself, who is a learned writer, can see this "nigger" only through Auld, who is the "other." And it is the "otherness" of Auld that is both repudiated and controlled by the narrator's balanced antithesis. By converting the otherness of Auld (and, consequently, his "nigger") into discourse, Douglass becomes the master of his own situation. And the white man, who wants a silently laboring brute, is finally (and ironically) visible to himself and a learned reading public only through the discourse of the articulated black spokesman.

Much of the remainder of the *Narrative* counterpoints the assumption of the white world that the slave is a brute against the slave's expanding awareness of language and its capacity to carry him toward new dimensions of experience. Chapter 7 (the one following the Auld encounter), for example, is devoted to Douglass's increasing command of the word. He discovers *The Columbian Orator,* with its striking messages of human dignity and freedom and its practical examples of the results of fine speaking. He also learns the significance of that all-important word *abolition.* Against these new perceptions, he juxtaposes the unthinking condition of slaves who have not yet acquired language skills equal to his own. At times he envies them, since they (like the "meanest reptile") are not fully and self-consciously aware of their situation. For the narrator, language brings the possibility of freedom but renders slavery intolerable. It gives rise to his decision to escape as soon as his age and the opportunity are appropriate. Meanwhile, he bides his time and perfects his writing, since (as he says in a telling act of autobiographical conflation) "I might have occasion to write my own pass."

Douglass's description of his reaction to ships on the Chesapeake illustrates that he did, effectively, write his own pass: "Those beautiful vessels, robed in purest white, so delightful to the eye of freemen, were to me so many shrouded ghosts to terrify and torment me with thoughts of my wretched condition." He continues with a passionate apostrophe that shows how dichotomous are his own condition and that of these white, "swift-winged angels."

> You are loosed from your moorings, and are free: I am fast in my chains, and am a slave! You move merrily before the gentle gale, and I sadly before the bloody whip! You are freedom's swift-winged angels, that fly around the world; I am confined in bands of iron! O that I were free! O, that I were on one of your gallant decks, and under your protecting wing! Alas! betwixt me and you, the turbid waters roll. Go on, go on. O that I could also go! Could I but swim! If I could fly! O, why was I born a man, of whom to make a brute! The glad ship is gone; she hides in the dim distance. I am left in the hottest hell of unending slavery. O God, save me! God, deliver me! Let me be free! Is there any God? Why am I a slave? I will run away. I will not stand it. Get caught, or get clear, I'll try it.

When clarified and understood through language, the deathly, terrifying nothingness around him reveals the grounds of being. Freedom, the ability to chose one's own direction, makes life beautiful and pure. Only the man free from bondage has a chance to obtain the farthest reaches of humanity. From what appears a blank and awesome backdrop, Douglass wrests significance. His subsequent progression through the roles of educated leader, freeman, abolitionist, and autobiographer marks his firm sense of being.

But while it is the fact that the ships are loosed from their moorings that intrigues the narrator, he also drives home their whiteness and places them in a Christian context. Here certain added difficulties for the black autobiographer reveal themselves. The acquisition of language, which leads to being, has ramifications that have been best stated by the West Indian novelist George Lamming, drawing on the relationship between Prospero and Caliban in *The Tempest:*

> Prospero has given Caliban Language; and with it an unstated history of consequences, an unknown history of future intentions. This gift of language meant not English, in particular, but speech and concept as a way, a method, a necessary avenue towards areas of the self which could not be reached in any other way. It is in this way, entirely Prospero's enterprise, which makes Caliban aware of possibilities. Therefore, all of Caliban's future—for future is the very name for possibilities—must derive from Prospero's experiment, which is also his risk.

Mr. Auld has seen that "learning" could lead to the restiveness of his slave. Neither he nor his representer, however, seem to understand that it might be possible to imprison the slave even more thoroughly in the way described by Lamming. The angelic Mrs. Auld, however, in accord with the evangelical codes of her era, has given Douglass the rudiments of a system that leads to intriguing restrictions. True, the slave can arrive at a sense of being only through language. But it is also true that, in Douglass's case, a conception of the preeminent form of being is conditioned by white, Christian standards.

To say this is not to charge him with treachery. Africa was for the black Southern slave an idealized backdrop, which failed to offer the immediate tangible means of his liberation. Moreover, whites continually sought to strip Africans of their distinctive cultural modes. Vassa's isolation and perplexity upon his arrival in the New World, which are recorded in a passage previously cited, give some notion of the results of this white offensive. Unable to transplant the institutions of his homeland in the soil of America—as the Puritans had done—the black slave had to seek means of survival and fulfillment on that middle ground where the European slave trade had deposited him. He had to seize whatever weapons came to hand in his struggle for self-definition. The range of instruments was limited. Evangelical Christians and committed abolitionists were the only discernible groups standing in the path of America's hypocrisy and inhumanity. The dictates of these groups, therefore, suggested a way beyond servitude. And these were the only signs and wonders in an environment where blacks were deemed animals, or "things." Determined to move beyond a subservient status, cut off from the alternatives held out to whites, endowed with the "feeling" that freedom is the natural condition of life, Douglass adopted a system of symbols that seemed to promise him an unbounded freedom. Having acquired language and a set of dictates that specified freedom and equality as norms, Douglass becomes more assured. His certainty is reflected by the roles he projects for himself in the latter part of his *Narrative*. They are all in harmony with a white, Christian, abolitionist framework.

During his year at Mr. Freeland's farm, for example, he spends much of his time "doing something that looked like bettering the condition of my race." His enterprise is a Sabbath school devoted to teaching his "loved fellow-slaves" so they will be able "to read the will of God." His efforts combine the philanthropic impulse of the eighteenth-century man of sympathy with a zeal akin to Jupiter Hammon's.

Having returned to Mr. Auld's house after an absence of three years, he undertakes a useful trade and earns the right to hire out his own time. All goes well until he attends a religious camp meeting one Saturday night and fails to pay the allotted weekly portion of his wages to his master. When Auld rebukes him, the demands of the "robber" are set against the natural right of a man to worship God freely. Once again, freedom is placed in a Christian context. Infuriated, Douglass decides that the time and circumstances are now right for his escape. When he arrives in New York, he feels like a man who has "escaped a den of hungry lions" (a kind of New World Daniel), and one of his first acts is to marry Anna Murray in a Christian ceremony presided over by the Reverend James W. C. Pennington. It would not be an overstatement to say that the liberated self portrayed by Douglass is firmly Christian, having adopted cherished values from the white world that held him in bondage. It is not surprising, therefore, to see the narrator moving rapidly into the ranks of the abolitionists—that body of men and women bent on putting America in harmony with its professed ideals. Nor is it striking that the *Narrative* concludes with an appendix in which the narrator justifies himself as a true Christian.

In recovering the details of his past, then, the autobiographer shows a progression from baffled and isolated existent to Christian abolitionist lecturer and writer. The self in the autobiographical moment (the present, the time in which the work is composed), however, seems unaware of the limitations that have accompanied this progress. Even though the writer seems to have been certain (given the cohesiveness of the *Narrative*) how he was going to picture his development and how the emergent self should appear to the reader, he seems to have suppressed the fact that one cannot transcend existence in a universe where there is *only* existence. One can realize one's humanity through "speech and concept," but one cannot distinguish the uniqueness of the self if the "avenue towards areas of the self" excludes rigorously individualizing definitions of a human, black identity.

Douglass grasps language in a Promethean act of will, but he leaves unexamined its potentially devastating effects. One reflection of his uncritical acceptance of the perspective made available by literacy is the *Narrative* itself, which was written at the urging of white abolitionists who had become the fugitive slave's employers. The work was written to prove that the narrator had indeed been a slave. And while autobiographical conventions forced him to portray as accurately as possible the existentiality of his original condition, the light of abolitionism is always implicitly present, guiding the narrator into calm, Christian, and publicly accessible harbors. The issue here is not simply one of intentionality (how the author wished his utterances to be taken). It is, rather, one that combines Douglass's understandable desire to keep his job with more complex considerations governing "privacy" as a philosophical concept.

Language, like other social institutions, is public; it is one of the surest means we have of communicating with the "other," the world outside ourselves. Moreover, since language seems to provide the principal way in which we conceptualize and convey anything (thoughts, feelings, sensations, and so forth), it is possible that no easily describable "private" domain exists. By adopting language as his instrument for extracting meaning from nothingness, being from existence, Douglass becomes a public figure.

He is comforted, but also restricted, by the system he adopts. The results are shown in the hierarchy of preferences that, finally, constitute value in the *Narrative*. The results are additionally demonstrated by those instances in the *Narrative* where the work's style is indistinguishable from that of the sentimental-romantic oratory and creative writing that marked the American nineteenth century.

Had there been a separate, written black language available, Douglass might have fared better. What is seminal to this discussion, however, is that the nature of the autobiographer's situation seemed to force him to move to a public version of the self—one molded by the values of white America. Thus Mr. Auld can be contained and controlled within the slave narrator's abolitionist discourse because Auld is a stock figure of such discourse. He is the penurious master corrupted by the soul-killing effects of slavery who appears in poetry, fiction, and polemics devoted to the abolitionist cause.

But the slave narrator must also accomplish the almost unthinkable (since thought and language are inseparable) task of transmuting an authentic, unwritten self—a self that exists outside the conventional literary discourse structures of a white reading public—into a literary representation. The simplest, and perhaps the most effective, way of proceeding is for the narrator to represent his "authentic" self as a figure embodying the public virtues and values esteemed by his intended audience. Once he has seized the public medium, the slave narrator can construct a public message, or massage, calculated to win approval for himself and (provided he has one) his cause. In the white abolitionist William Lloyd Garrison's preface to Douglass's *Narrative,* for example, the slave narrator is elaborately praised for his seemingly godlike movement "into the field of public usefulness." Garrison writes of his own reaction to Douglass's first abolitionist lecture to a white audience:

> I shall never forget his first speech at the convention—the extraordinary emotion it excited in my own mind—the powerful impression it created upon a crowded auditory, completely taken by surprise—the applause which followed from the beginning to the end of his felicitous remarks. I think I never hated slavery so intensely as at that moment; certainly my perception of the enormous outrage which is inflicted by it, on the godlike nature of its victims, was rendered far more clear than ever. There stood one, in physical proportion and stature commanding and exact—in intellect richly endowed—in natural eloquence a prodigy—in soul manifestly "created but a little lower than the angels"—trembling for his safety, hardly daring to believe that on the American soil, a single white person could be found who would befriend him at

all hazards, for the love of God and humanity. Capable of high attainments as an intellectual and moral being—needing nothing but a comparatively small amount of cultivation to make him an ornament to society and a blessing to his race—by the law of the land, by the voice of the people, by the terms of the slave code, he was only a piece of property, a beast of burden, a chattel personal, nevertheless!

Obviously, a talented, heroic, and richly endowed figure such as Garrison describes here was of inestimable "public usefulness" to the abolitionist crusade. And the Nantucket Convention of 1841 where Garrison first heard Douglass speak may be compared to a communicative context in which the sender and receiver employed a common channel (i.e., the English language) to arrive at, or to reinforce for each other, an agreed-upon message. Douglass transmits the "heroic fugitive" message to an abolitionist audience that has made such a figure part of his conceptual, linguistic, and rhetorical repertoire.

The issue that such an "autobiographical" act raises for the literary analyst is that of authenticity. Where, for example, in Douglass's *Narrative* does a prototypical black American self reside? What are the distinctive narrative elements that combine to form a representation of this self? In light of the foregoing discussion, it seems that such elements would be located in those episodes and passages of the *Narrative* that chronicle the struggle for literacy. For once literacy has been achieved, the black self, even as represented in the *Narrative,* begins to distance itself from the domain of experience constituted by the oral-aural community of the slave quarters (e.g., the remarks comparing fellow slaves to the meanest reptiles). The voice of the unwritten self, once it is subjected to the linguistic codes, literary conventions, and audience expectations of a literate population, is perhaps never again the authentic voice of black American slavery. It is, rather, the voice of a self transformed by an autobiographical act into a sharer in the general public discourse about slavery.

How much of the lived (as opposed to the represented) slave experience is lost in this transformation depends upon the keenness of the narrator's skill in confronting both the freedom and the limitations resulting from his literacy in Prospero's tongue. By the conclusion of Douglass's *Narrative,* the represented self seems to have left the quarters almost entirely behind. The self that appears in the work's closing moments is that of a public spokesman, talking about slavery to a Nantucket convention of whites:

> While attending an anti-slavery convention at Nantucket, on the 11th of August, 1841, I felt strongly moved to speak, and was at the same time much urged to do so by Mr. William C. Coffin, a gentleman who had heard me speak in the colored people's meeting at New Bedford. It was a severe cross, and I took it up reluctantly. The truth was, I felt myself a slave, and the idea of speaking to white people weighed me down. I spoke but a few moments, when I felt a degree of freedom, and said what I desired with considerable ease. From that time until now, I have been engaged in pleading the cause of my brethren—with what success, and with what devotion, I leave to those acquainted with my labors to decide.

The Christian imagery ("a severe cross"), strained reluctance to speak before whites, discovered ease of eloquence, and public-spirited devotion to the cause of his brethren that appear in this passage are all in keeping with the image of the publicly useful and ideal fugitive captured in Garrison's preface. Immediately before telling the reader of his address to the Nantucket convention, Douglass notes that "he had not long been a reader of the 'Liberator' [Garrison's abolitionist newspaper]" before he got "a pretty correct idea of the principles, measures and spirit of the anti-slavery reform"; he adds that he "took right hold of the cause . . . and never felt happier than when in an anti-slavery meeting." This suggests to me that the communication between Douglass and Garrison begins long before their face-to-face encounter at Nantucket, with the fugitive slave's culling from the white publisher's newspaper those virtues and values esteemed by abolitionist readers. The fugitive's voice is further refined by his attendance and speeches at the "colored people's meeting at New Bedford," and it finally achieves its emotionally stirring participation in the white world of public discourse at the 1841 Nantucket convention.

Of course, there are tangible reasons within the historical (as opposed to the autobiographical) domain for the image that Douglass projects. The feeling of larger goals shared with a white majority culture has always been present among blacks. We need only turn back to the earlier discussion of Hammon, Wheatley, and Vassa to see this. From at least the third decade of the nineteenth century this feeling of a common pursuit was reinforced by men like Garrison and Wendall Phillips, by constitutional amendments, civil rights legislation, and perennial assurances that the white man's dream is the black man's as well. Furthermore, what better support for this assumption of commonality could Douglass find than in his own palpable achievements in American society?

When he revised his original *Narrative* for the third time, therefore, in 1893, the work that resulted represented the conclusion of a process that began for Douglass at the home of Hugh Auld. The *Life and Times of Frederick Douglass Written by Himself* is public, rooted in the language of its time, and considerably less existential in tone than the 1845 *Narrative*. What we have is a verbose and somewhat hackneyed story of a life, written by a man of achievement. The white externality has been transformed into a world where sterling deeds by blacks are possible. Douglass describes his visit to the home of his former master who, forty years after the slave's escape, now rests on his deathbed:

> On reaching the house I was met by Mr. Wm. H. Buff, a son-in-law of Capt. Auld, and Mrs. Louisa Buff, his daughter, and was conducted to the bedroom of Capt. Auld. We addressed each other simultaneously, he called me "Marshal Douglass," and I, as I had always called him, "Captain Auld." Hearing myself called by him "Marshal Douglass," I instantly broke up the formal nature of the meeting by saying, "not *Marshal,* but Frederick to you as formerly." We shook hands cordially and in the act of doing so, he, having been long stricken with palsy, shed tears as men thus afflicted will do when excited by any deep emotion. The sight of him, the changes which time had wrought in him, his tremulous hands constantly in motion, and all the circumstances of his condition affected me deeply, and for a time choked my voice and made me speechless.

A nearly tearful silence by the black "Marshal" (a term repeated three times in very brief space) of the District of Columbia as he gazes with sympathy on the body of his former master—this is a great distance, to be sure, from the aggressive young slave who appropriated language in order to do battle with the masters.

A further instance of Douglass's revised perspective is provided by his return to the home plantation of Colonel Lloyd on the Wye River in Talbot County, Maryland:

> Speaking of this desire of mine [to revisit the Lloyd Plantation] last winter, to Hon. John L. Thomas, the efficient collector at the Port of Baltimore, and a leading Republican of the State of Maryland, he urged me very much to go, and added that he often took a trip to the Eastern Shore in his revenue cutter *Guthrie* (otherwise known in time of war as the *Ewing*), and would be much pleased to have me accompany him on one of these trips. . . . In four hours after leaving Baltimore we were anchored in the river off the Lloyd estate, and from the deck of our vessel I saw once more the stately chimneys of the grand old mansion which I had last seen from the deck of the *Sally Lloyd* when a boy. I left there as a slave, and returned as a freeman; I left there unknown to the outside world, and returned well known; I left there on a freight boat and returned on a revenue cutter; I left on a vessel belonging to Col. Edward Lloyd, and returned on one belonging to the United States.

The "stately chimneys of the grand old mansion" sounds very much like the Plantation Tradition, and how different the purpose of the balanced antithesis is in this passage from that noted in the delineation of the slave's realization of language as a key to freedom ("What he most dreaded, that I most desired"). This passage also stands in marked contrast to the description of ships on the Chesapeake cited earlier ("those beautiful vessels . . . so many shrouded ghosts"). The venerable status of the *Guthrie* is now matched by the eminence of the marshal of the District of Columbia.

Douglass, in his public role, often resembles the courteous and gentlemanly narrator of Vassa's work—a man determined to put readers at ease by assuring them of his accomplishments (and the sterling company he keeps) in language that is careful not to offend readers' various sensibilities. It is strikingly coincidental that the *Life and Times of Frederick Douglass* was reprinted in 1895, the year in which its author died and Booker T. Washington emerged as one of the most influential black public spokesmen America had ever known.

The Problematic of Self in Autobiography: The Example of the Slave Narrative

Annette Niemtzow

> *But I am called Soaphead Church. I cannot remember how or why I got the name. What makes one name more a person than another? Is the name the real thing, then?*
>
> <div align="right">Toni Morrison, <i>The Bluest Eye</i></div>

Slave narratives suggest an ur-moment in history—the moment when a person who has existed in a previously oral world picks up an implement for the first time to communicate through writing. Each of us who can now read and write has recapitulated this process in our own lives, shifting as we did from preliteracy to literacy. Thus, all autobiographies, as part of the community of the literate, have also made this journey. Yet, in western autobiographical writing, the child's passage from darkness to light is not ordinarily described in terms of a movement from the oral to the written world. But the slave narrative has chosen its own conception as one of its subjects. Print on the page—itself often treated by writers as the neutral in the tension between *récit* and *discours*—provides joy in the slave's text, for it is itself the *récit* of the text, and its creation is at least part of the *discours*. Therefore, in addition to the political and historical demands slave narratives make of us, they merit the most serious literary attention of the student of the theory of

---

From *The Art of Slave Narrative: Original Essays in Criticism and Theory,* edited by John Sekora and Darwin D. Turner. © 1982 by Western Illinois University.

autobiography because of the knots of primary critical problems they set before us.

What moves a person to write about his or her life? Or more particularly, what moved black men and women, recently free from the atrocity of slavery, to write about their experiences? Ostensibly, the primary motivation was to woo white readers to hatred of slavery and to love of abolition; many narratives were written at the urging of white sympathizers. But white encouragement and a sense of moral responsibility do not provide sufficient motivation. There were other motivations too, motivations the slave writer shares with all other autobiographers, who attempt to describe a self—no matter how painful to acknowledge—before it disappears, to describe a self which, no matter how despicable, is so fragile that unless it exists on paper, it will hardly exist at all. The slave narrative suggests a paradox: the slave, happily ceasing to be a slave, describes his or her slave self to preserve it just as it is about to cease to be. Slave narratives, like all other autobiographies, involve nostalgia through memory. And nostalgia, in the slave autobiographer's case, involves the slave in a contradictory act of recapturing a self that the slave wishes to cast off, and also involves the slave in a search for a self capable of blooming into an admirable adult, an adult who is a writer. The slave narrative, like Wordsworth's *Prelude,* though more covertly, must trace the growth of the writer's mind.

There are at least three conditions needed for all autobiography, including slave narratives: the history of an individual; an interest in the content as well as the form of that person's life; and an implicit identity between the writer and the protagonist. Yet, it is this last requirement that is most challenged by the slave narrative, which by its very existence underscores the gap between "I" as writer and "I" as protagonist, so much so that the writer often cries out, like Frederick Douglass at the recollection of the moment of literacy:

> And as I read, behold! the very discontent so graphically
> predicted . . . had already come upon me. I was no longer
> the lighthearted, gleesome boy, full of mirth and play, that
> I was when I landed in Baltimore.

Douglass's outburst is at once a lament and a cheer. The boy without letters is not merely a loathsome self to be rid of (Douglass writes on, "I often found myself regretting my own existence, and wishing myself dead"), but—and this truth becomes stronger and stronger in his revision—Douglass the boy, not Douglass the slave boy, has admirable traits toward which the adult writer feels positively. The boy will be the writer. Yet, the earlier self vanishes before his and our eyes; and the slave struggles to push it off at the very moment that the autobiographer reaches for it. The struggle is often so intense that the reader, black or white, cries out, in a confusion that suggests the power of the narratives, "How could this person have been a slave?" By "person," we mean the autobiographer who confusingly bears the name of the slave. Our cry is both one of sympathy against the institution of slavery as we acknowledge the achievement of the autobiographer and a cry of disbelief in the possibility that a slave could become the autobiographer. Indeed, such disbelief has traditionally been interpreted as simple racism and, indeed, it has had grave historical consequences. For example, historians, following Ullrich Phillips's position, have argued that slave narratives could not be treated as historical documents, since they were so clearly edited beyond recognizability by white abolitionists. But, more theoretically, the reader's disbelief is a tribute to the author of the slave narrative, for the doubting reader is inevitably at one with the doubting writer. The memory fades of a child forced, as Douglass describes his early self, to live like a pig; and we all struggle to connect the slave boy to the skilled narrator Douglass has shaped. Yet, this connection is crucial for the writer and the reader, or the text dissolves as autobiography; factuality is one of the necessities of the genre. As Georges Gusdorf has written:

> Autobiography becomes possible only under certain metaphysical preconditions. To begin with, at the cost of a cultural revolution humanity must have emerged from the mythic framework of traditional teachings and have entered into the perilous domain of history. The man who takes the trouble to tell of himself knows that the present differs from the past and that it will not be repeated in the future; he has become more aware of differences than of similarities; given the constant change, given the uncertainty of events and of men, he believes it a useful and valuable thing to fix his own image so that he can be certain it will not disappear like all things in this world.

Slave autobiographers were probably less certain than most autobiographers that history would "not be repeated in the future," but they worked to make it so. For slave narratives had a deep social mission which would insure that the future would not repeat the past, and that was to establish the identity of each slave as slave no longer, but sentient, intelligent human being. I read, I write, therefore I am, says the slave autobiographer of nineteenth-century America. And the message was private as well as public.

But in order to receive the pledge of selfhood available through autobiography, the slave first had to conquer still another immediate enemy, beyond the master or any one else in the white world. Having unraveled the puzzles of signs and riddles of language, the former slave had yet to face the blank page. This, of course, was not a problem unique to the slave. For almost a hundred years, Henry James's description of the condition of the nineteenth-century American writer has stood unchallenged, accepted as bland fact; the writer has been portrayed as a man facing the "blankness" of the American circumstance. And while James's term echoes with much truth, it is probably more interesting as an example of a nineteenth-century white male writer's *sense* of his own circumstance than it is an analysis of the reality of the writer's condition. James's description has obscured recognition of the substantial aspects and qualities of American life which provided these writers with much more than "blankness." For example, by the time of Emerson's "Nature," there were almost two centuries of Puritan ontology available as legacy which by the nineteenth century developed into a full-blown sense of national mission that the nation repeatedly imposed on all white men, including its writers.

But what was the condition of the slave man or the slave woman? Houston Baker has pointed out that the slave condition was indeed the essence of "blankness":

> [For slaves there were] scarcely any a priori assumptions to act as stays for self-definition. [The slave] was a man of the diaspora, a displaced person imprisoned by an inhumane system. . . . For [him] the white externality provided no ontological or ideological certainties; in fact, it explicitly denied slaves the grounds of being. . . . Instead of the ebullient sense of a new land offering limitless possibilities, the slave staring into the heart of whiteness around him, must have felt as though he had been flung into existence without a human purpose. The white externality must

have loomed like the Heidiggerian "nothingness," the negative foundation of being.

But if Baker is right, and it would seem he is, what a circumstance we must posit for the slave autobiographer—a nonexisting self trying to outline its existence on a blank sheet of paper. Yet, nothing, we know, can come of nothing. Thus, if we agree that the slave's self does not exist at the moment before the conception of the slave narrative, we must postulate the fullness of the previously described "blank" page at the moment of that conception. Is there evidence for this hypothesis? Let us consider the case of Frederick Douglass's *Narrative* in its 1845 edition.

From its beginning, Douglass's narrative suggests how full the page he faced, indeed, was. He seems to understand from the start that he lacks the very stuff on which autobiography must consist:

> I have no accurate knowledge of my age, never having seen any authentic record containing it.

When he proclaims that only a "record" can provide "authentic" information, he forms an allegiance to knowledge through the written word, which the rest of the text sustains. He struggles to define himself by the very terms established not only by white culture, but by autobiography itself.

> My mother was named Harriet Bailey. She was the daughter of Isaac and Betsey Bailey, both colored and quite dark. My mother was of darker complexion than either my grandmother or grandfather.
>
> My father was a white man. He was admitted to be such by all I ever heard speak of my parentage.

Douglass is unhappily forced back to the oral tradition. Yet, he provides his autobiography with the requisite parents, even if the required father can only be produced within a bitterly ironic frame. Such conjoinment unites Douglass biologically and literally to the white world, but who, after all, wants to claim such a father? And yet, Douglass must, if he is to come closer to legitimating his own self. For what Douglass intuits is that autobiography, more dogmatically perhaps than society itself, requires a child to have a father, if he or she is to have a self. An autobiographer, even a slave autobiographer or even a fictive autobiographer, must pay attention to the ideal of father, if the reader is to believe in the existence of that self. So a Nick Carraway or a Robinson Crusoe must tell us about his father. So too, inevitably, the slave must create a father that is important and, therefore, white.

I am, of course, not suggesting that a great many slave fathers were not white men, who used slave women as sexual chattel, laborers in the primal sense, but what I am suggesting (though I am reticent to do so lest I seem to absolve white men of their well-earned guilt) is that the autobiographer's proclaimed link to the white father has other sources too. And autobiography itself—a form by which one proclaims a self—is a crucial and neglected source. Fifty years before Douglass's work, Benjamin Franklin provided a model which solidified the idea that autobiographies were expected to bow to paternity by offering an account which made all history coexist with the history of his family. The past is his ancestry; the present is himself; the future is his son. Yet, like Douglass, Franklin works to establish that history, for his access to it was limited by a familial, indeed, cultural shift from an oral and, therefore, untrustworthy tradition to a world of stable "notes," of incontrovertible written documents. And his relationship to his father, as he was well aware, was not as positive as his assertion that he was the "youngest Son of the youngest Son for 5 Generations back" implies. The position of youngest sons in English and American families was tentative at best, and in Franklin's instance, his "Bookish inclination" seems to have been his father's despair. And yet, Franklin's subjection to fictions proves successful; his fictions themselves fulfill their pledge. Through words—the source of his potential disinheritance—his position as youngest son—comes the means of his social reinstatement. Only through print, by correcting what he typographically calls "errata," does he develop the necessary paternal lineage. His genealogy, approached through fictive glasses, grants him grace. In a text which purports to be enamored of American values, Franklin manages to establish a legacy from a distinguished English line. He manages even to claim superiority from his maternal line in a text establishing male lineage. In other words, through writing, Franklin turns all situations to the service of his self. Through the written word, he gains connections which he could not otherwise claim. For example, by verbal association, he becomes indirectly honored by Cotton Mather, whom he claims has mentioned one of his ancestors in his *Magnalia*. Later, to make himself Mather's legatee, at least by implication, he renames the latter's *Bonifacius, An Essay upon the Good* as "Essays to do Good," so that Mather's work becomes a precursor of his own. Franklin's *Autobiography* then points out the origins of its form, with its insistence on paternity and family, deriving, as it must have, from the need of society to identify a person through public means. The oral formulae of genealogy available in the "begats" of biblical and epic worlds suggest primitive patriarchal ways to define a self, ways which, in fact, lasted. But it is Franklin, white printer, who controls in the end what is real. He establishes patriarchal lineage as the norm for written as well as oral forms of autobiography.

What happens when Douglas tries to engage in an act comparable to Franklin's? He can claim only a paternity which humiliates him, although it links him to past white culture, with which, by necessity, he must connect. What the comparison between Franklin and Douglass most underscores is an odd circumstance of autobiography, that, in order to write about a self that is socially acceptable in print, one must already have a self which is, at least in part, acceptable in print. Since the required description of self depends upon a legacy from the past, one must seem to have the same. But Douglass is in a position of constructing a self from an ontological and historical void. Yet, his is a blankness which he does not have to confront in print if he assents, as he does, to describe himself through the white norm (it is as if he proclaims, "I have parents too," in answer to a white challenge inherent in the form). As a result, he experiences the void, not as a void at all, but as a space occupied by white culture. So he says of his own ignorance of his age, "By far the larger part of the slaves know as little of their age as horses know of theirs, and it is the wish of most masters within my knowledge to keep their slaves thus ignorant." Whenever he acknowledges that there is a void—that he lacks self-definition—he finds that the white man has filled it, with a definition of blacks—as animals or as "niggers." Douglass is not angry with the fact that the void is filled, but with *how* it is filled. Indeed, at first, Douglass the boy accepts the white view of his self as animal and tries to eradicate his skin—by the white account—the determinant of an animal self:

> I was probably between seven and eight years old when I left Colonel Lloyd's plantation. I left it with joy. . . . I spent the most part of . . .[the next] three days in the creek, washing off the plantation scurf, and preparing myself for my departure.
> The pride of appearance which this would indicate was

> not my own. I spent the time in washing, not so much because I wished to, but because Mrs. Lucretia had told me I must get all the dead skin off my feet and knees before I could go to Baltimore; for the people in Baltimore were very cleanly, and would laugh at me if I looked dirty. . . . It was almost a sufficient motive, not only to make me take off what would be called by pig-drovers the mange, but the skin itself.

The scene is crucial—a symbolic shedding of one self for another. Yet, the new definition, like its predecessor, is external. Douglass, the boy, accepts a definition of human selfhood which a white woman has given him, and which includes cleanness or, to his mind, whiteness as a high value; his own skin then becomes the connection to pigs, and he wages war against the skin itself—as if it, rather than the white definition, were the cause of his woe. Douglass describes himself as a child who thinks that finding his self will involve shedding his past—losing all links to blackness (presumably Africa itself was the most sizeable enemy), but because of this belief, he also forsakes familial and domestic ties, the ones which later, as autobiographer, he will feel it necessary to reestablish. It is autobiography itself which forces him to apologize for the boy who so readily desired to lose his skin *and* his home:

> The ties that ordinarily bind children to their homes were all suspended in my case. I found no severe trial in my departure. My home was charmless; it was not home to me; on parting from it, I could not feel that I was leaving any thing which I could have enjoyed by staying.

Douglass feels apologetic for his own feelings, for autobiography has presented him with contradictory demands—that he shed his juvenile slave self and that he be sentimental about that self and its setting, the home. Not only in its particular form in the nineteenth century, when Wordsworth urged autobiography toward nostalgia, but also, even at its most general, autobiography is a form set in nostalgia. Once again, Douglass finds his own experience wanting when measured against a white autobiographical norm, and yet, it is to that norm that he has assented by accepting the act of autobiography itself. By choosing to write, at a time when most blacks, still slaves in America, were not literate, he has offered a move of assent toward structuring a self for white readers. Indeed, by agreeing to face the page at all, by agreeing to the primacy of the definition of a self as someone who reads and writes, Douglass has adopted a white definition of selfhood, and tries to attain it.

Given the importance then for Douglass of literacy as the first and primary act of selfhood, the achievement of literacy naturally assumed a central moment in his autobiography—as it did in other black autobiographies as well. The impetus toward literacy occurred when Douglass learned that, for whites, humanness depended on literacy. Thus, for him, the recognition that reading is an entry to freedom is not a self-conceived notion, but, like cleanliness, a standard defined by whites. Douglass conceives of a self which he will form in opposition to his master's wishes, but ironically he forms it within his master's rules. Reading, like autobiography itself, in the nineteenth century provides an entry to the kind of self acceptable to white culture.

Douglass's autobiography then, by virtue of its genre, unconsciously pays tribute to a definition of self created by whites. As he says about speaking to whites, "The truth was, I felt myself a slave, and the idea of speaking to white people weighed me down. I spoke but a few moments when I felt a degree of freedom, and said what I desired with considerable ease." It is his ability to communicate with whites—an act which involves learning a white ideolect—which permits him to move from a position as slave to a position as a self. The autobiography itself then—the act of writing itself—helps him to this self, even if it is, as it must be within the terms of the form, only a self defined by whites; all definitions of self are defined by whites, for the word itself posits a concept controlled by whites. Thus, it is appropriate that a black man chooses a poem by a white man to name Douglass:

> The name given me by my mother was "Frederick Augustus Washington Bailey." I, however, had dispensed with the two middle names long before I left Maryland so that I was generally known by the name of "Frederick Bailey." I started from Baltimore bearing the name of "Stanley." When I got to New York, I again changed my name to "Frederick Johnson," and thought that would be the last change. But when I got to New Bedford, I found it necessary again to change my name. The reason of this neces-

> sity was, that there were so many Johnsons in New Bedford, it was already quite difficult to distinguish between them. I gave Mr. Johnson the privilege of choosing me a name, but told him he must not take from me the name of "Frederick." I must hold on to that, to preserve a sense of my identity. Mr. Johnson had just been reading the "Lady of the Lake," and at once suggested that my name be "Douglass." From that time until now I have been called "Frederick Douglass;" and as I am more widely known by that name than by either of the others, I shall continue to use it as my own.

This is potentially a sad moment, when Douglass might recognize how random his identity has been; indeed, he might recognize that white men dominate the world of language, so that, while it is a black man, a "Mr. Johnson" that names him, "Mr. Johnson" finds a creditable name only in a text written by a white poet. White men simply have a prior claim on valid names. Douglass has no legitimacy, as it were, when he is named by his mother—the black woman. He must be christened via access to white language, for it is the white man who controls all things, especially language. To acquire a self, he sacrifices his tie to a past which does not provide access to the white world.

In specific, from his original position as a man unable to meet the expectations of autobiography at the beginning of his narrative, Douglass invents (in the Renaissance sense) a self recognizable to himself and to his white readers by the end of the tale. He is reborn as if he were Benjamin Franklin himself:

> In the afternoon of the day when I reached New Bedford, I visited the wharves, to take a view of the shipping. Here I found myself surrounded with the strongest proofs of wealth. . . . From the wharves I strolled around and over the town, gazing with wonder and admiration at the splendid churches, beautiful dwellings, and finely-cultivated gardens.

The scene evokes memories of Franklin's entry onto Market Street Wharf in Philadelphia, where he walks about, conscious of everyone else's wealth and his own impecunious state. Douglass becomes Franklin recast, in his own, but yet familiar, terms. Indeed, Douglass's idea of self would have met Franklin's approval:

I found employment, the third day after my arrival, in stowing a sloop with a load of oil. It was new, dirty, and hard work for me; but I went at it with a glad heart and a willing hand.

Most slave narratives embrace a work ethic. Indeed, as if to meet Franklin's approval, as his narrative draws near its end, Douglass not only has a job, but a wife as well, and certainly in terms that are historically explicable, he offers even a copy of his marriage certificate in the text—the evidence that he has met the terms of a society that lives by the written word. Even in the 1845 version of his autobiography, Douglass seizes a self which is defined by a white code; he meets the terms of the American male's dream in white, conventional limits—with a job and a wife.

Lest though we be too hard on Douglass, we need to remember again what little choice he had and how well his artfulness served him as writer. Yet, he described the only self the white world thought possible, though there are hints too, in the autobiography that Douglass glimpsed other, indeed, original, ideas of self, which were not open to him within the terms permitted by the genre. The nate lesson is discernible in his experience with another slave, Sandy Jenkins, to whom he tells the story of the violence of his present master, Mr. Covey; Sandy urges him to carry a *root* which would prevent any white man from whipping him, though Douglass, with his insistence that the *word* is the key to freedom, with his belief in rationality and his rejection of magic—implicit in his African inheritance, through his mother, scoffs at the idea, but (or so he maintains) "To please him, I at length took the root, and according to his direction, carried it upon my right side." To Douglass's surprise, Covey does not strike him on his return, but the day is Sunday. Yet, Monday, the "root was fully tested." As Covey attacks him, Douglass says, "I resolved to fight" and did. Douglass recognizes the importance of the *roots* incident and, even in his skepticism, acknowledges that Sandy, the traditions of Africa, the brotherhood between the two men though slaves, reverberates for his future: "My long-crushed spirit rose, cowardice departed, bold defiance took its place; and I now resolved that, however long I might remain a slave in form, the day had passed forever when I could be a slave in fact." And yet, Douglass's does not carry Sandy's *roots* message of a true self possible through an African and possibly maternal lineage with him to a new sense of self and to a new form of writing; instead, despite hesitations which he has about the highly structured, class society he enters, autobiography itself—the very language and terms of being which it embraces—define the man we call "Frederick Douglass."

As we might suspect, this need to censure any self foreign to the dominant sociolect is not Douglass's problem alone; it exists for all minority autobiographers. In fact, though offering an extreme instance, Douglass was caught inside a problem inherent to literacy, which only "the strongest" writers—to use Harold Bloom's term—manage to elude, as best they can. I have read; I can and will write; therefore, I am, might be a more accurate, revised form of any autobiographer's resolve, even if they must reshape their life experiences sharply to match the terms of the genre. Writers show originality as they can. Douglass's *roots* incident shows this truth; so too does the kind of play against genre in which many slave autobiographers engage. For example, Lunsford Lane wittily opened his narrative of 1842:

> On the 30th of May, 1803, I was ushered into the world; but I did not begin to see the rising of its dark clouds, nor fancy how they might be broken and dispersed until some time afterward.

Lane is acknowledging the required beginning of an autobiography ("Give origins"); yet, he offers the passive "I was ushered" as a clever substitute for mother and father, though his text soon returns to more conventional realities, after looking at the clouds. The autobiographer may also attempt to elude the requirements of the genre by changing genres. Linda Brent's *Incidents in the Life of a Slave Girl* suggests how autobiography failed to meet the female slave's attempt to express her self or to discover her self. Brent, in fact, found the genre so wanting that she fled it—and ran straight to the arms of the more usuable, more female domestic novel.

Male slave narratives, indeed male autobiographies, are frequently stories of triumph in a public sphere. Culturally, the male weighs his worth against public criteria; females do not—in general. As a result, female autobiographies, including female slave narratives, often show greater freedom than male works against the confines of the genre. For example, Brent opens more positively than Douglass, talking about a self which is her own and which is, therefore, protected, although an external definition—the word "slave"—tries to subsume it: "I was born a slave," she writes, "but I never knew it till six years of childhood had passed away." Brent seems to experience feelings which are prior to the external definition of self, by white society or by a white genre. Even her record of her family is not shaped by external requirements; for her, the father is not necessarily the central figure in self-definition. Brent mentions a father and a mother, but then she claims a brother—and in an act foreign to white male autobiography—a grandmother. To make autobiography bend to the requirements of the black family—particularly to find space for the often important role of the grandmother—has been the achievement of the modern black autobiographer, especially the female autobiographer, but Brent pointed the way. Unlike Douglass, she does not apologize for what she lacks. Her mother's death, for instance, enters the narrative, not only because the story demands it, but because the death is significant to Brent's personal history. Her mother dies, and then, asserts Brent, "for the first time, I learned by the talk around me, that I was a slave." Her story is more independent of the formulations imposed by autobiography than Douglass's, or to put the case differently, since the narrative is *occupied,* so to speak, by a self, there is less void for a predetermined form to fill up. What this analysis suggests is a simple formula for autobiography, which can be applied almost mathematically: as the self increases, the impact of autobiography as a genre decreases. The self, not the form, begins to control. Since the greatest art requires the most control, it is not surprising, despite the literary achievement of some slave narratives, that the slave, hardly in a position to offer a formed self, able to take charge, did not universally achieve great art. Slave autobiographies succeed artistically when they show such control—note Douglass's gift for imagery. Is Brent's then the great achievement of slave narratives? She did, after all, come to the page with some self intact. The answer, of course, is no. Ultimately, she lacked the means to assert a viable self which is independent of strictures imposed on her by the nineteenth-century definitions by gender. To own a definition of self consonant with one's feeling, a definition not susceptible to the vagaries of form and society might well serve as an existential definition of freedom, but such a self was not possible for Brent to maintain.

Male slaves needed names and control of naming to achieve adult identity; for women, bound to give up their names, the issues of adulthood were different. Maturity for a nineteenth-century woman required three victories: control of her own chastity; attainment of a successful marriage; and ability at marriage and mothering of her own children. Only those women who achieved success in these three areas—that is, women who remained virgins until their triumph as wives and mothers—were enviable women, female adults. As a slave, from the start, Brent possessed no opportunities for these victories. She could not be a successful woman, in white terms, any more than Douglass could be a successful man; yet, like him, she would not acknowledge her a priori failure. In the same way that Douglass as slave is deprived of a manhood, defined as the ability to choose work and have a wife, Brent is deprived of true womanhood. And, like Douglass's, Brent's narrative centers around her attempt to gain the virtues of adulthood, as it is defined by the white culture. Yet, to gain these virtues demanded by successful womanhood, Brent must permit her own voice, the authentic voice which often violates the terms of the prescribed norm, to be consumed—in her case by the work which described perfect womanhood, the domestic novel. Ironically then, Brent's flight from autobiography leads her to a new prison.

The domestic novel was the form which articulated the issues of female chastity and marital compliance. Richardson's *Pamela* and *Clarissa,* emulated again and again in eighteenth-century American texts, provided the model for the mythologizing of a shared female dilemma. What Franklin's *Autobiography* offered and took from male writers, these fictionalized accounts of women's lives—fictionalized autobiographies or biographies—offered female writers: a highly rationalized form of a deeply felt situation. The unhappy irony is that this "female" form was in fact the work of a white man. Like Franklin's mastery of the world, Pamela's happiness in the reformed household of Mr. B. offered a confidence that the problems of femaleness would be answered by the inevitability of the solution, which itself was guaranteed by the wholeness of literary form; the sense of completion of the story suggests a resolution for the life—a self confirmed, so much in Pamela's own case that she was able to act, even on others. Clarissa could withstand Lovelace's sexual attacks. The female crisis is temporary—answered, even if only answered, by Clarissa's death and self-defined retention of her own chastity, secure, despite outside verities. The confidence then of the domestic novel mocked the condition of the female slave, and it is therefore not surprising that the domestic novel swallows Linda Brent's voice.

Why then did she choose the domestic novel as a means of ordering her autobiography? The answer is, I think, simple. She gains much by making the form her own. First of all, the form is hers, by right of the traditions of gender, and in accepting it, she places herself in the community of women so well defined in the nineteenth century, and this, despite the fact, that the form originates with males. Secondly, the form itself gives her what the culture has taken away—a means to assert her own righteousness and, in female terms, class. By writing in the form of the domestic novel, she, a black woman, asserts her desire to be chaste; thus, she establishes her kinship with other women, white and black. At once, she is like them. And, as a female author—whose rank she joins at once by her choice, she gains too a reader—a female reader, whose existence she alludes to again and again. These are no small gains, for the isolated slave woman, without a sure peer group—a slave woman almost alone, isolated partly by her ability to read and write.

But Brent gains and loses more too by her use of the domestic novel. First, she gains access to the right to record what was clearly for her a central experience, that which defined her sense of self, that which distinguished her plight as a female slave from that of the male: the constant fear of rape. And yet, if the domestic novel opens up sexual harassment as a possible subject for a story—indeed, a story which black and white women can share, the terms of the discussion in the novel transform a story of victim and rapist into a story of a young lady faced with a Byronic seducer. The clear loss in this choice of story suggests that all—and perhaps all retellings—originate with men. Possibly, despite the hopes of critics like Cixous or Irigaray, there is no "écriture feminine." As Brent tells the story, her master "came [to see her] with frowning brows, that showed a dissatisfied state of mind":

> He ordered me to stand before him. I obeyed. "I command you," said he, "to tell me whether the father of your child is white or black." I hesitated. "Answer me this instant!" he exclaimed. I did answer. He sprang upon me like a wolf, and grabbed my arm as if he would have broken it. "Do you love him?" said he, in a hissing tone.

A violent, potentially mortal encounter is reduced, not heightened, by the stock animal imagery Brent offers. Her owner, Dr. Flint, continues—by her account—to speak with words which seem the hyperbole of a betrayed and wounded lover, not the physical threat they undoubtedly were:

> "You turn aside all my good intentions toward you. I don't know what it is that keeps me from killing you."

Even his final threat and his assertion of his control of her very being is lost in the intensity of a seduction romance: "You are my slave, and shall always be my slave," says Dr. Flint. Brent cannot write openly about her own experience, for the domestic novel requires that its author, like its intended readers, be above offering a frank discussion of sexuality, much less of the distortion of sexuality in rape.

The rape of slave women is notably absent from the 6,000 extant collected slave narratives by women and men. Instead, there exist only polite stories of impolite "seductions," like Brent's by Dr. Flint. For example, in William Wells Brown's narrative of 1847, he recounts the story of a slaveowner named Walker who offered a "vile proposal" to a quadroon named Cynthia, but a more graphic account of rape or forced sexuality, comparable to accounts of floggings, is missing from his and other narratives. During the American Freedmen's Inquiry Commission Interviews of 1863 and in speeches by slaves, the rhetoric about the sexual harrassment of the female slave is heightened—to meet the rhetorical requirements of the genres involved; yet, these too, for reasons of politeness inherent in the genres as they exist in our society, lack the immediate graphic intensity of descriptions of other violence.

Rape was not a legal reality for the slave woman in nineteenth-century America, but when she began to read and then to write, she discovered that it was not a literary reality either. Indeed, the slave woman whose very circumstance must have made her painfully aware of the double burden under which she lived was now forced to eliminate that for which she suffered—her gender itself. Details of Brent's narrative suggest, for example, that she was a woman with sexual experiences of her own. She does, after all, without the circumstances entering the text, birth two children with a man she cares for, but she cannot tell the story of her sexuality for she must work to establish that she is a chaste woman, pursued by the devil. The domestic novel—with its polite sense of female sexuality—forbids her her own life as subject.

The images of the domestic novel seem to mesmerize Brent so

that she seems unable to grasp the miracle—at the end of the text—
of her own escape from slavery to freedom. Consider, for instance,
the penultimate paragraph of her narrative:

> Reader, my story ends with freedom, not in the usual way,
> with marriage. I and my children are now free! We are as
> free from the power of slaveholders as are the white people
> of the north; and though that, according to my ideas, is
> not saying a great deal, it is a vast improvement in *my* con-
> dition. The dream of my life is not yet realized. I do not
> sit with my children in a home of my own.

Brent apologizes for having *only* freedom—technically, we might say
that she disparages one genre, the slave narrative, for one she values
more, the domestic novel; but as a woman, not a writer conscious
that she is trapped between genres, she experiences only pain, not
irony—pain at her failure to attain that toward which the style of her
narrative has led her: "Reader," she says, "my story ends with freedom, not in the usual way, with marriage." As she says, her dream
is not yet fulfilled; she does not have a home (a hearthstone) and,
need we add, a husband. The requirements for a content self, even
the requisite for any self—freedom—are trivialized by the fictive
forms offered the female slave.

Current debates about autobiography often center on questions
of whether autobiography is or is not fiction. Delighted by the discovery that autobiographers use techniques of fiction, critics have
managed to confuse themselves about the nature of the genre itself.
Autobiographies have shape; this does not mean that autobiographies are fiction. Autobiography is a *genre,* a term which suggests
that a literary work belongs to a group of like works which can be
classified together. And autobiography, like the novel, is a problematic genre. It minimally requires that a self try to recount some facts
of the life he or she has led. Although autobiography uses fictive
techniques, it is not false; it is fictive, not faithless. Its truth is incomplete, as any vision must be—for it is mediated by genre which
brings with it literary techniques. Genre itself has consequences for
what we learn of truth. Slave narratives are one of autobiography's
own children. Their distortions of truth are attributable to their origins. In the nineteenth century, the white man owned language. He
also owned form. And so, and this should not be surprising, the
page was full, as the black writer faced it. In the nineteenth century,

the black writer struggled to subvert white forms. In the twentieth century, the black writer has continued that struggle, both by subversion and by creating new forms.

Ellison's Invisible Man in the hospital of "Liberty," the most American of paint factories, experiences the archetypal moment of a challenge to his identity, when he wakes up to the sight of people who demand of him again and again, "WHAT IS YOUR NAME?" and, then, "WHO . . . ARE . . . YOU?" But he, at least, recognizes the problem. As he says, "I realized that I no longer knew my name" and "Left alone, I lay fretting over my identity." The slaves' literary space was more public, as they accepted public definitions of their selves. They did not know the truth which the Invisible Man proclaims, "When I discover who I am, I'll be free." The slave narrative was a step toward freedom; it, as a caged form, was not the journey.

Language in Slavery

Ann Kibbey

For historians who use slave narratives to document the immediate physical and social facts of slave life, Frederick Douglass's *Narrative of the Life of Frederick Douglass, an American Slave, Written by Himself* offers a frustratingly low yield. Beside Solomon Northup's detailed account of living quarters, diet, work life, holidays, and family relations, Douglass's *Narrative* must seem spare, incomplete, even misleading in its portrayal of the slave experience—an incendiary polemic written more to fuel the abolitionist cause than to convey the nature of the slave experience. To read the *Narrative* from this point of view, however, is to misapprehend how Douglass's text treats slavery and to be needlessly disappointed. Unlike Northup, Douglass focuses on the linguistic significance of bondage: He tersely portrays masters and slaves almost solely in terms of their linguistic acts because, for him, the reality of slavery is a profoundly rhetorical one. He charts his own relentless progress to freedom as the acquisition of an ever deeper understanding of language use in a slave economy, and the realization of his own freedom at the Nantucket antislavery convention is preeminently a linguistic event. Douglass's perspective is an important one, for as sociolinguists have discovered, "peoples do not all everywhere use language to the same degree, in the same situations, or for the same things. . . . Languages, like other cultural traits, will be found to vary in the degree and nature of their integra-

From *Prospects: The Annual of American Cultural Studies* 8 (1983). © 1983 by Cambridge University Press.

tion into the societies and cultures in which they occur." Douglass was acutely sensitive to the linguistic system of slave society, of the ways in which language was used—and withheld—by one human being to enslave another.

The last event described in Douglass's text, an antislavery convention at Nantucket in 1841, allows us to locate the narrator as a particular person speaking at a particular historical place and time, for it is the point where the narrator and persona of Douglass converge. According to William Lloyd Garrison's preface, Douglass, in his "first speech at the convention . . . proceeded to narrate some of the facts in his own history as a slave, and in the course of his speech gave utterance to many noble thoughts and thrilling reflections." The speech Douglass delivered to his Nantucket audience was presumably a version of the 1845 *Narrative,* and the convention thus identifies the social perspective from which to understand the narrative the reader has just finished. The narrator explains that he had successfully escaped from slavery economically in 1838, settling in New Bedford, where he became a "working man." But although the status of free laborer was a necessary condition of his freedom, only the meetings of the antislavery movement offered him the full liberty he desired. Drawn to the movement by Garrison's newspaper, *The Liberator,* Douglass began to attend its meetings in New Bedford. Despite his freedman's status, the system of slavery still influenced him, for he recalls, "I seldom had much to say at the meetings," and he tells us why: "The truth was, I felt myself a slave, and the idea of speaking to white people weighed me down." Finally, at the Nantucket convention, Douglass left slavery behind linguistically, as well as economically: "I felt strongly moved to speak, and was at the same time much urged to do so by Mr. William C. Coffin, a gentleman who had heard me speak at the colored people's meeting at New Bedford. It was a severe cross, and I took it up reluctantly." Douglass presents this moment of reluctance as his last slavish act: "I spoke but a few moments, when I felt a degree of freedom, and said what I desired with considerable ease. From that time until now, I have been engaged in pleading the cause of my brethren." Collapsing the subsequent four years into this last phrase, the narrator thus identifies himself to the reader at the conclusion of his text.

It would be wrongheaded to suppose that Douglass defines his freedom here in terms of acceptance into white society. The Nantucket meeting was an event whose racial character was finally *not* definitive for him—this is what he discovered when he refused to heed his feelings of reluctance. Beginning with his speech at Nantucket, it no longer mattered whether he spoke to whites or blacks; he would say the same thing to either.

In defining his freedom, Douglass directs our attention instead to the significance of his economic and social condition. Indeed, throughout his narrative Douglass never fails to describe the circumstances of his linguistic discoveries, carefully detailing his moves through all the economic gradations of slavery as he traces his linguistic progress to Nantucket. Whatever temptation we may have to cast his self-presentation in the heroic proportions of the extraordinary individual—as Garrison does in his preface—there is nothing in the *Narrative* to suggest that Douglass ever felt himself especially gifted in his insights. On the contrary, it was largely a matter of finding himself in the social conditions conducive to such insights. Midway through the *Narrative,* when he calls attention to himself as a free author, he also makes us acutely aware of the fortuitousness of his present status:

> It is possible, and even quite probable, that but for the mere circumstances of being removed from that plantation to Baltimore, I should have to-day, instead of being here seated by my own table, in the enjoyment of freedom and the happiness of home, writing this Narrative, been confined to the galling chains of slavery.

It was not simply that the Auld house in Baltimore offered more opportunities for escape than Colonel Lloyd's plantation. Rather, the move was a change of status and circumstance—from potential field hand in a rural society to a quasi-personal servant in the economically various world of urban life. Douglass describes at length the variety of people he met in Baltimore and his new experiences at the Auld house, explaining how this new situation changed his thinking. Similarly, Douglass discovered his freedom through his participation in the social reality of the Nantucket convention, as a liberty achieved at a particular place and time, within a particular economic and social structure.

The way Douglass describes his own experience implies that the narrative voice in this text is at once an individual voice and a social product, a subject that is intersubjective in its very constitution. Douglass expresses this same paradoxically social idea of the individual in the structure of this text, for it is impossible to find the *Narrative*'s thread of continuity by recourse to his life alone. The narrator repeatedly interrupts his personal story—most emphatically by refusing to relate the means of his escape. Granted that Douglass's personal history is told more fully than anyone else's, it still does not exclusively determine the meaning or purpose of the text. Both topically and structurally, then, the narrator assumes that his use of words is a social act, that his own words are the product of social relationships. The narrator comprehends the meaning of words as a social relation *between* people, not as the creation of an isolable private consciousness. Indeed, there is no such thing as thinking outside an orientation toward possible expression. The individual consciousness is not the source of explanation; rather, it is, itself, in need of explanation.

The implications of this view become more apparent when we consider Douglass's description of a situation antithetical to his own at Nantucket: Colonel Lloyd's slaves singing on their way to Great House Farm. In the temporary freedom of the woods, where presumably they could sing as they pleased, their spontaneous songs were nearly unintelligible to the child Douglass who overheard them. Douglass briefly quotes some words about the master's farm and the slaves' journey to it, telling us that "into all of their songs they would manage to weave something of the Great House Farm," but the rest "to many would seem unmeaning jargon." Although "full of meaning to themselves," these other words meant nothing to anyone else. Only the most obedient slaves on the outlying farm were chosen for this jaunt to the home plantation to pick up monthly provisions. Slaves competed for the distinction because it was "associated in their minds with greatness. . . . They regarded it as evidence of great confidence reposed in them by their overseers . . . a high privilege, one worth careful living for." These slaves lived materially and socially at the pleasure of their master, as the purpose of their journey suggests. With the center of authority in slavery as their destination both literally and linguistically, they had lost the power to communicate any meaning but the slaveholder's. What meaning of his own the slave might express remained trapped in a private semantics: He sang "wild songs," Douglass says, with the same sorrow as a "man cast away upon a desolate island." For the slave who had accommodated his inner world to the possibilities of expression in a slave society, the songs symbolized the contradiction in the word "slave": the human being who was nonhuman, the social being who was nonsocial.

Unlike Douglass at Nantucket, whose spontaneous speech engaged him in society and whose articulate sense of self was inherently social, the slave bound for the home plantation sang precisely *because* there was no one to hear him—or no one who, under the system of slavery, would have been recognized as human. For those who had won the overseer's approval—who had assented to a system in which slaves were socially identifiable only in terms of their masters—the meaning they shared among themselves had become nonexistent to others, even to another slave like Douglass overhearing them. Douglass says of himself as hearer, "I did not, when a slave, understand the deep meaning of those rude and apparently incoherent songs. I was myself within the circle; so that I neither saw nor heard as those without might see and hear." From the perspective of Nantucket, Douglass sees the paradox of his own response, both that he did not understand the "deep meaning" of what he heard and that "to those songs I trace my first glimmering conception of the dehumanizing character of slavery." His own contradictory doubleness matched what he heard, a rude and incoherent jargon that was, nevertheless, profoundly expressive: "I have sometimes thought that the mere hearing of those songs would do more to impress some minds with the horrible character of slavery, than the reading of whole volumes of philosophy on the subject could do." What Douglass knows discursively, as a free author outside the circle of slavery, is what he intuitively sensed within: The meaningless jargon itself symbolized a capacity and intent to create meaning that could not be eradicated even in slavery's most obedient victims. Even a slave who was willing to please his master could not cease to be human.

Nor is it clear, finally, that this was even what the master demanded of the slave. Through his rhetorical form, Douglass evokes the sense in which contradiction pervaded the slaves' way of life: "They would sometimes sing the most pathetic sentiment in the most rapturous tone, and the most rapturous sentiment in the most pathetic tone." This crisscross rhetoric conveys the idea of a parallelism, an equivalence, deliberately violated; a contradiction forcibly introduced; a paradox that need not be there. The rhetorical form suggests the full social significance of these songs as expressions of the slave's condition: The negation of the real stature of the slave as a human subject depended on the reality of its opposite—the affirmation of a reciprocal relation between master and slave as two equally human beings. Although the ideology of slavery defined the master-slave relation as radically nonreciprocal, the economy of slavery depended on the humanity of slaves in order to function, for the daily work of running a plantation could not be carried on without basic intelligence and linguistic proficiency in the slave. Although racist justifications sought some middle ground between human and animal for the slave, Douglass argues there was no such middle ground for the slave in fact: The master demanded that the slave be the social equivalent of a rhetorical crisscross, at once human and nonhuman, to labor within this social structure.

The rhetorical figure in his description of the slaves' song is part of a larger argument Douglass makes throughout his *Narrative* in showing what constitutes the enslavement of a person. Someone born into the economic condition of a slave is not thereby born into a slave's linguistic system. Each economic slave must be linguistically enslaved, must have the paradox of "slave" forced on him, must be coerced to discover and enter a structure of meaning that denies his humanity. Douglass begins his narrative with the recollection of his own linguistic enslavement. Its purpose was to consign him to a wholly "natural" world without social dimensions, a world that ideally would cancel his linguistic ability to name and develop primary human relations. Like other slaves, he did not know his age: "I do not remember to have ever met a slave who could tell of his birthday. They seldom come nearer to it than planting-time, harvest-time, cherry-time, spring-time, or fall-time." In slavery's division of humanity according to the arbitrary categories of "nature" and "culture," inherent in "slave" and "master," the slave reckoned his life by the natural cycle of the seasons, without the distinct sense of personal history the cultural knowledge of age confers. As a child, Douglass felt the contrast but felt powerless against it: "The white children could tell their ages. I could not tell why I ought to be deprived of the same privilege. I was not allowed to make any inquiries of my master concerning it. He deemed all such inquiries on the part of the slave improper and impertinent."

Similarly, Douglass did not know who his father was, beyond knowing he was white, because, as he explains, "the means of knowing was withheld from me." For interracial slaves such as Douglass, the patriarchal structure of the slave system was sustained, paradoxically, by the absolute denial of paternity. The master-father's relation to his child was purely economic: The master had authority over his slave children because he owned their mothers, not because he was their father. Douglass knew the identity of his mother, but here "mother" functioned only as a signal with an invariable meaning. She was admitted to be his mother only to place her child within the primary dichotomy of master and slave. Socially, Harriet Bailey seems to have meant very little to her son. Douglass explains:

> My mother and I were separated when I was but an infant—before I knew her as my mother. It is a common custom, in the part of Maryland from which I ran away, to part children from their mothers at a very early age. . . . I never saw my mother, to know her as such, more than four or five times in my life.

Hired out to another farm, Harriet Bailey had to travel twelve miles on foot at night to see her son. Not surprisingly, Douglass describes his mother as someone who was, for the most part, absent from his life: "I do not recollect of ever seeing my mother by the light of day. She was with me in the night. She would lie down with me, and get me to sleep, but long before I waked she was gone." What little he knows of her is followed by a summary account of her death, and in his final words about her, Douglass describes himself as one who had succumbed to slavery's refusal to acknowledge any societal existence for the slave:

> Very little communication ever took place between us. . . . I was not allowed to be present during her illness, at her death, or burial. She was gone long before I knew anything about it. Never having enjoyed, to any considerable extent, her soothing presence, her tender and watchful care, I received the tidings of her death with much the same emotions I should have probably felt at the death of a stranger.

His mother had become any one of the indefinite number of other people whose interchangeability was a function of the master's equation of all slaves as chattels.

Douglass concludes this chapter with his recollection of the whipping of his Aunt Hester, remembering this event as the one that sealed his enslavement: "It struck me with awful force. It was the blood-stained gate, the entrance to the hell of slavery, through which

I was about to pass." He saw his aunt suspended from a joist in the kitchen and heard her screams as he watched the master beat her hideously bloody. He narrates the material and physical details with vivid immediacy, bringing his description to focus on the master's whip and the slave's blood: "After rolling up his sleeves he commenced to lay on the heavy cowskin, and soon the warm, red blood (amid heart-rending shrieks from her, and horrid oaths from him) came dripping to the floor." Here Douglass saw for the first time the violence of slavery, and his response was the reaction of a person whose self-conception had been circumscribed by the word "slave": "I was so terrified and horror-stricken at the sight, that I hid myself in a closet, and dared not venture out till long after the bloody transaction was over. I expected it would be my turn next." He imagined himself as the master's next victim, but there was no ostensible reason to believe that he would likewise be punished, for her crime could not have been his. Supposedly, she was beaten because she had been absent when her master had wanted his mistress, and worse still, she had been found in the company of a male slave on a neighboring plantation. The boy's terrified response to the whipping was that of the slave who had learned what "female relative" meant in the language of slavery: He understood her slave condition—and thereby his own—as the rationalization for the master's gratuitous violence.

The language of slavery affected Douglass's earliest experiences of social relations by making the members of his family socially indistinguishable, erasing their individual differences and their distinctive relation to himself. To the extent that enslavement succeeded, he was left without any social orientation but the name of "slave." In the language of slavery, Douglass's father was not his father, and his mother, despite her efforts, was not his mother. The separation of slave and master amounted to a system of double referentiality, where the meanings of social names such as "father" and "mother" were always referred to the primary terms "master" and "slave" for their meaning. For the slave, this resulted in a cancellation of signification altogether, for it eliminated the variability of meaning, the capacity of a word to vary with context that distinguishes a word from a mere signal.

The elimination of family terms was only the most salient example of the more general tendency in the language of slavery to destroy the meaning of all words except "master" and "slave." The meaning of other words was always a function of these two social

meaning of other words was always a function of these two social names, this single, invariable social context, for the master could interrupt any discourse by invoking this privileged context at any time. The scene with Aunt Hester is again suggestive, for although Douglass gives the ostensible reason for her whipping, he also remarks that the master "would at times seem to take great pleasure in whipping a slave." As a slave child, Douglass responded not to any moral system of punishment but only to the master's prerogative of inflicting violence on his slaves. The possibility that makeshift reasons could always be found for whipping becomes far more explicit in the description of the slaves responsible for the care of the master's horses, who also "never knew when they were safe from punishment." Quoting the master's irate condemnation of his slaves, Douglass exposes the absurdity of supposing that whipping was somehow connected with rationality or justice:

> It was painful to stand near the stable-door, and hear the various complaints against the keepers when a horse was taken out for use. "This horse has not had proper attention. He has not been sufficiently rubbed and curried, or he has not been properly fed; he was too hot or too cold; he had too much hay, and not enough grain; or he had too much grain, and not enough hay. . . ." To all these complaints, no matter how unjust, the slave must answer never a word.

This is hardly the representation of a single, typical speech. Rather, it is an elision of all the master's speeches—and the slaves' responding silences—that rhetorically illustrates the principles of language use in slavery. The isolation of these words from their differing nonlinguistic situations makes them referentially meaningless. Or rather, they refer not to the care of horses but to the power of the master over the slave, the power to maim and kill. Even the signification of words that would rationalize this brutality as "punishment" is sacrificed to the definition of "master" and "slave." One suspects that such solipsists as Colonel Lloyd preferred not to have to use words at all—to use nonverbal signals instead. Resorting to "all kinds of stratagems" to prevent his slaves from stealing fruit from his garden, the colonel finally hit on the "most successful" solution of tarring the fence around the garden. The slaves quickly learned the colonel's private signal code, that tar on their person meant a whipping: "It was deemed sufficient proof that he had either been into the garden, or had tried to get in. In either case, he was severely whipped by the chief gardener."

Throughout the narrative, Douglass characterizes mastery as the power to inflict injury at will. Such cruelty may seem to be only the perverse luxury of the masters and thus ancillary to the system of slavery itself, but the relationship between overseer and slave suggests instead that such violence was intrinsic to language use in slavery. Linguistically, Douglass gives us a double view of the overseer. On the one hand, he seems strikingly nonverbal or, at least, like Colonel Lloyd, nonsensical. Douglass writes of Mr. Severe, "from the rising till the going down of the sun, he was cursing, raving, cutting, and slashing among the slaves of the field." The slaves preferred his successor, Mr. Hopkins, because "he made less noise." Mr. Covey crudely enacted the bestial image of "the snake": "He seldom approached the spot where we were at work openly, if he could do it secretly. . . . When we were at work in the cornfield, he would sometimes crawl on his hands and knees to avoid detection, and all at once he would arise nearly in our midst and scream out." These overseers seem to have communicated little more than the sensation of threat and the reality of their whips. But on the other hand, the overseers could not always have sounded like raving "noise" to the slaves and still run the work of the plantation. When the overseer's words concerned the nature of work to be done, these words had to be meaningful to the laborer. Douglass describes at some length the linguistic virtuosity of a highly successful overseer, the exemplary Mr. Gore who was promoted to overseer of the home plantation. Gore was "a man possessing, in an eminent degree, all those traits of character indispensable to what is called a first-rate overseer." Gore never indulged in the kinds of verbal exchange that mitigated the power relationship between overseer and slave. He only spoke to the slaves when he gave orders: "Mr. Gore was a grave man, and though a young man, he indulged in no jokes, said no funny words, seldom smiled. . . . Overseers will sometimes indulge in a witty word, even with the slaves; not so with Mr. Gore. He spoke but to command, and commanded but to be obeyed."

Linguistically, Gore's relation to the slaves was defined by his commands. But these commands were not only imperatives; they were also self-justifying declarations of his own perfection, of his right to autocratic authority:

There must be no answering back to him; no explanation was allowed a slave, showing himself to have been wrongfully accused. Mr. Gore acted fully up to the maxim laid down by the slaveholders,—"It is better that a dozen slaves suffer under the lash, than that the overseer should be convicted, in the presence of the slaves, of having been at fault."

The authority to command carried with it the authority to accuse—in effect, the authority to name, for as Douglass explains, "to be accused was to be convicted . . . the one always followed the other with immutable certainty." In the ideology of the slaveholders, masters but not slaves created meaning. To put it another way, Gore wished his utterances to be performative, automatically to produce in social actuality what he declared in speech. The overseer might create the illusion of a performative utterance, but it could never be more than that. His words were not performative but only imperative, requiring a silent response of obedience from the hearer to give them the quality of a command. Concealed in the slave's silence was the unacknowledged linguistic reciprocity between speaker and hearer. Gore gave his words meaning with his whip, for only the lash—or the threat of it—could make the actions around him resemble what he had commanded. He seems to have sensed better than most overseers the intrinsic relationship between referential meaning and whipping: "He dealt sparingly with his words, and bountifully with his whip, never using the former where the latter would do as well. When he whipped, he seemed to do so from a sense of duty." His whip expressed, as it were, the overseer's intent and capacity to create meaning for his words. But while the noise of the whip seemed to be the sound of autonomous validation, it was actually the validating silence of the slave that gave the overseer's words the social reality of a command.

The situation of work continually belied the master-slave relationship for what it really was: a relationship between two persons, an intersubjective creation of meaning. That creation of meaning could not belong solely to the master, much as he desired to make it seem so; but for the slave who exposed Gore's autocratic illusion, who exposed the overseer's vulnerability to the slave's cooperation, the punishment was death: When the slave Demby plunged into a creek to escape a whipping and refused Gore's command to come out, the overseer summarily shot and killed him, murdering the human counterevidence to the overseer's power of declaring social reality. Once Demby had proved he was irreducibly another person, distinct from his master's mental fantasy of a "slave," he had, in Gore's terms, become "unmanageable." By placing himself beyond the reach of the overseer's whip and then refusing to obey a command, Demby had made his own silence audible, revealing the fragility of referential meaning in the overseer's words.

Whether obedient or disobedient, the slave laborer was always in a contradictory situation. Whipped to repudiate his subjectivity by affirming the social structure that denied it, the laborer, nevertheless, asserted his subjectivity by his very response to being whipped. The overseer seemed to be, by contrast, a model of logic: Gore was "always at his post, never inconsistent. . . . He was, in a word, a man of the most inflexible firmness and stone-like coolness." But even the overseer who thoroughly believed the ideology that empowered him could not thereby escape its contradictions. With crisscross rhetoric, Douglass satirizes Gore's presumption of faultless logic and complete autonomy: "He was just the man for such a place, and it was just the place for such a man. . . . His words were in perfect keeping with his looks, and his looks were in perfect keeping with his words." The empty tautology reverberates in the complete enclosure of this rhetorical figure, as the seeming logic of fitness is undermined by the reversal of words. Drawing a circle of isolation rhetorically again, Douglass presents the reduction of a human being to an invariable context. But this time there is a satiric parallelism (the repetition of "perfect keeping"), implying that this reduction was far more complete than the slave's. Certainly the slaves who worked under this overseer were acutely aware that he lacked their social knowledge, that he did not perceive or understand either his humanity or theirs: "He was, of all the overseers, the most dreaded by the slaves. His presence was painful; his eye flashed confusion."

When Gore murdered Demby, "a thrill of horror flashed through every soul upon the plantation, excepting Mr. Gore. He alone seemed cool and collected." Unlike the slave, Gore lived in an isolation of his own making, produced by his belief in the master's idea of a slave. Although the overseer spoke in society, it was not a society he could be aware of, as such, and still retain his power. He was oblivious of the real linguistic and social relationship between overseer and slave, to the acts of silent reciprocity that maintained him in his position of command. From the perspective of his imaginary, self-created autonomy, he addressed a social void, and his acts of violence showed the depth of his desire to believe that there was no other human being out there.

Although Douglass portrays owners as well as overseers in acts of gross brutality, most of his descriptions of violence focus on overseers. The overseer, lacking the authority conferred by ownership, had to establish his mastery through his commands and his violence if he was to establish it at all. For the master who actually owned slaves, however, the slave market offered the inestimable advantage of a completely nonlinguistic, nonviolent definition of the relationship between master and slave. It offered the opportunity to assert his mastery through the pseudo language of money, a medium much better suited to express the master's belief that slaves were ideally a repetition of identical forms.

Douglass himself was never actually sold on the slave market, but he did face the prospect of sale after his first owner's death. Sent back from Baltimore to the home plantation to be "valued with the other property" in the inventory of his master's estate, Douglass found himself "ranked with horses, sheep, and swine" in the juxtaposition of human and nonhuman property: "There were horses and men, cattle and women, pigs and children, all holding the same rank in the scale of being and . . . all subjected to the same narrow examination . . . the same indelicate inspection." Certainly the gross mixture of human beings and animals implied the degradation of the slaves into bestiality, but there was more to the inventory than this. The slaves were not only ranked with animals in the "scale of being"; they were also priced for potential sale as commodities on the market. The inventory was a uniquely revealing event for Douglass: "At this moment, I saw more clearly than ever the brutalizing effects of slavery upon both slave and slaveholder." Where relatively kind treatment at the Auld house in Baltimore had engendered the illusion that the word "slave" marked a social status, however low, the inventory disclosed his real status as a chattel, as alienable property outside the social hierarchy altogether.

The rationale for the "division" of slaves was the "valuation" that preceded it; that is, the division of the estate reconstituted the social world of the slave according to the purely economic concept of the slave as a priced object of property. But because the slaves remained human nonetheless, they experienced the inventory as the imposition of a new kind of social relationship. As Douglass says, "I had now a new conception of my degraded condition." The inventory produced a "new conception" of slavery because it redefined the slave purely in terms of exchange value in the marketplace. The inventory was the moment when the idea of "slave" was transposed from a linguistic system to a monetary system of exchange.

The concept of the commodity in the capitalist marketplace is similar to the concept of the slave in the language of slavery, in that both conceal the same kind of contradiction. In the capitalist market, although the rationale for exchange presupposes that commodities have a particular use in the concrete world, a use dependent on their individual qualities, exchange value expresses only the abstract character of objects, giving them the uniform status that makes equivalence in exchange possible. The exchange value makes an object a "social hieroglyphic" because the commodity as a single entity conceals a contradiction between its use value (its inherent particularity), on the one hand, and its abstract, arbitrary exchange value, on the other. Money, the medium of exchange, completes this mystification because it expresses only the exchange value. That is, the system of exchange functions by ignoring the qualitative difference among commodities, by "translating" all objects into the same medium, money.

Analogously, the master's language use in slavery—insofar as it achieved its purpose—concealed the qualitative differences among slaves, giving them a uniform status in which their equivalence was expressed without recognizing their inherent particular qualities or their particular histories. The humanity of the slave—precisely what made him economically useful to the slaveholder—was concealed in the idea of the nonhuman "slave," the abstract idea that justified the master's ownership. This congruity of contradiction made it possible for owners to reinterpret their slaves as commodities without changing anything. What the capitalist market offers that language does not is the additional mystification of money. The slave market symbolically exploited the categories of the capitalist market to divest the slave of any trace of individual and historical particularity, for money expresses only an abstract, invariable identity. Money itself is defined by the complete absence of particularity; its concrete, qualitative aspects, as well as its similarity to other objects, must be absolutely irrelevant in order for it to function in its abstract character as the universal equivalent for exchange. To put this another way, the social character of money is invariable, for money "means" the same thing in every economic transaction—rather like Colonel Lloyd's tar on the garden fence. Thus money can never express the concrete particularity of an object or a person; it can only express the abstract ratios of labor time. This is all that a price "says," and Douglass's lack of concern with the actual prices ascribed to the slaves in their "valuation" evinces his sense that it was the attribute of price, rather than any specific numerical figure, that exclusively distinguished the slave in the social world. To assign a slave the attribute of price was more than a refusal to name his humanity, for the system of prices, unlike a linguistic system, precludes the possibility of naming the uniqueness of what is priced.

Thus the attribute of price not only equated the slave with objects sold in the market; it reified his being in a way that actively denied his individual humanity. Like the word "slave," money is an ideal form; but unlike the word that named the slave, price functioned to exclude the slave from any linguistic system, defining him with a signal that invariably symbolized his exclusion from language. As a person whose only social meaning was a price, the slave could never realize his subjectivity among others. Expressible only as money, his humanity could not be articulated. The human being at the foundation of the slave system, he was paradoxically not there. The slave market directly functioned to mystify the subjectivity, the human condition, of the participants in exchange because each slave was degraded to the status of a thing, a mere object mediating the relation between his seller and purchaser. The slave commodity was indeed a "social hieroglyphic" but far more profoundly than the commodity Marx describes. When the slaveholder asserted his social status by symbolically exploiting the market relations of free society, he negated the basis of language between himself and the slave. In the absence of language, the attribute of price signaled this negation.

It may seem that the categories of the market existed only at the boundary of slavery, without affecting the lives of slaves who were not sold. But to remind the slave of his potential attribute of price could effectively extend the owner's authority as the market defined it. For Douglass himself it was the experience of the inventory, short of being sold, that produced the "new conception of [his] degraded condition." The threat of the slave market could influence slaves to act as if the social relations between masters in the market had already redefined their status. For example, Douglass describes two bands of slaves who experienced their relations to each other only through their relations to their masters:

> When Colonel Lloyd's slaves met the slaves of Jacob Jepson, they seldom parted without a quarrel about their masters; Colonel Lloyd's slaves contending that he was the richest, and Mr. Jepson's slaves that he was the smartest, and most of a man. Colonel Lloyd's slaves would boast of his ability to buy and sell Jacob Jepson. Mr. Jepson's slaves would boast his ability to whip Colonel Lloyd. These quarrels would almost always end in a fight between the parties, and those that whipped were supposed to have gained the point at issue. They seemed to think that the greatness of their masters was transferable to themselves. It was considered as being bad enough to be a slave; but to be a poor man's slave was deemed a disgrace indeed!

In the slaves' hostility to each other, they experienced their social identity as a function of their masters' supposed attributes, as if they already had the status of commodities on the market. Their quarrels developed through the representation in Lloyd and Jepson of two different modes of mastery: one the power to buy and sell, the other the power to whip. The significance of the ensuing fight mixes both kinds of power: The outcome of their physical violence determines who is the "poor man's slave."

The fear of the slave market is also apparent in the apocryphal tale of the slave who did not recognize Colonel Lloyd on the road, and thus responded freely to his master's queries. Objecting to the treatment he received from his master, the slave naively continued on his way without ever realizing to whom he had spoken. Several weeks later he was shocked to discover that he was to be sold. Douglass draws the moral that this was the consequence of telling the "plain truth," but he tells the tale in the larger context of explaining just how wealthy Colonel Lloyd was. The outcome of the story also implies that the slave had misunderstood the nature of his anonymity. His real anonymity was that of a slave commodity, and if he ever presumed that the slave market was distant or unreal, he was simply deluding himself: Colonel Lloyd was wealthy enough to sell at once anyone who offended him.

I mean to suggest not that Douglass was a Marxist or that the *Narrative* is merely an illustration of Marxist theory but that Douglass's text is informed by a similar concern to understand the value of labor, the concept of private property, and the principles of the capitalist market. And beyond this, while Douglass's *Narrative* is by no means as systematic, thorough, or analytical in its treatment of these subjects, his ideas raise important questions about the slave economy that Marx does not discuss. Because Douglass wrote from the perspective of a slave laborer as well as a free laborer, his text requires an economic interpretation that fully accounts for what it meant to appropriate the principles of the capitalist market to buy and sell human beings.

What Douglass has to say about the slave economy he says within the frame of distinct narrative sequences. His treatment of Christianity is quite different, informing the *Narrative* at many points. His critique of American Protestantism goes well beyond denunciations of hypocritical Christians. Douglass opposes Christian idealism for asking the slave, in effect, to tolerate slavery by maintaining the slaveholder's illusion. Both the rhetoric and the form of the *Narrative* criticize Protestant idealism for ultimately consigning the slave to a silence equivalent to the silence demanded by the overseer and the isolated silence of the slave commodity.

Douglass uses Christian rhetoric ironically throughout his narrative. For example, in recounting an experience that can be considered a conversion of sorts, his discovery of the word "abolition," he adopts the phrases of religious conversion, but with an ironic undertone that announces their inappropriateness. In language reminiscent of the Christian's yearning to hear the Word of God, he says he "always drew near when that word was spoken." But "that word" for him is "abolition." "The light broke in upon me by degrees," he remembers, borrowing a favorite phrase of the Christian convert; but the light for Douglass is not the Divine Light: It is the human knowledge—gleaned from the stevedores at the Baltimore docks—that "abolition" denoted a social reality he might actually experience, not just the theoretical negation of his current condition. The Nantucket convention can also be seen as a religious event, a kind of Quaker meeting in which Douglass felt himself moved by the Inner Light to take on the "severe cross" of bearing witness; but again the religious allusions are ironic. The enthralling charm of ascesis, of suffering for a Christian purpose, dissipates with the actuality of his first words; and he testifies not at all of the deity, but of his own humanity instead.

The rejection of Christian conversion in Douglass's irony is part of his broader criticism of idealist thought. Although Douglass owed his initial conception of freedom to idealist antislavery tracts, he also became aware that his own dilemma could never be resolved at the level of language alone. Indeed, the pain of realizing the disparity between his real social condition and the ideal of freedom was torturous:

> As I writhed under it, I would at times feel that learning to read had been a curse rather than a blessing. It had given me a view of my wretched condition, without the remedy. It opened my eyes to the horrible pit, but to no ladder upon which to get out. . . . It was this everlasting thinking of my condition that tormented me. . . . I often found myself regretting my own existence, and wishing myself dead.

In the anguish of the entrapment Douglass was silent. Socially, and linguistically, these tracts showed him nothing more than a way of tolerating the master's system of double referentiality. He learned to speak without speaking and hear without hearing, to be the human being who was not there. As Scripture does for the Christian, the tracts articulated his conceptions for him: "They gave tongue to interesting thoughts of my own soul, which had frequently flashed through my mind, and died away for want of utterance." But although he now knew the words, he could not speak them or realize from his own experience what they meant. In a similar metonymic displacement, the world around him symbolically spoke of the freedom that was not his:

> [Freedom] was heard in every sound, and seen in everything. It was ever present to torment me with a sense of my wretched condition. I saw nothing without seeing it, I heard nothing without hearing it, and felt nothing without feeling it. It looked from every star, it smiled in every calm, breathed in every wind, and moved in every storm.

In this figure of personification Douglass had no voice, for his commitment to the ideal of freedom had alienated him from his own material and social existence—a transcendent experience, perhaps, but not one that *literally* allowed him to be free of slavery. Freedom existed only in his figurative imagination, and the world figuratively "spoke" of freedom for him only because, as a slave, he did not literally speak of it himself. While the pathetic fallacy he "heard" implied that freedom, not slavery, was his "natural" condition, it implied as well that only the nonhuman, literally speechless, world around him "thought" so. Douglass is ultimately critical of such idealism because, while it introduced him to the radical contradiction between slavery and freedom, it also implied that he could never expect to escape slavery except in thought.

Douglass would feel the same sense of impossibility years later, when as a field hand he spent Sunday in quasi-religious meditation inspired by ships on the Chesapeake. The white sails "were to me so many shrouded ghosts, to terrify and torment me with thoughts of my wretched condition." Freedom seemed impossibly out of reach, but "my thoughts would compel utterance," he explains, "and there, with no audience but the Almighty, I would pour out my soul's complaint, in my rude way, with an apostrophe to the moving multitude of ships." This is the only time Douglass names the rhetorical figure he uses, and the speech to the "Almighty" that he proceeds to quote is not "rude" at all but a tendentious and highly artificial rhetoric that is totally out of character for the narrator—the empty, formal voice of a slave who may have been present to God but was absent to himself. Returning us to the narrator's perspective, Douglass explains that he had addressed neither God nor the ships, but only himself: "Thus I used to speak to myself; goaded almost to madness at one moment, and at the next reconciling myself to my wretched lot." The tensions generated by idealism compelled him toward a raging lunacy, on the one hand, and an equally terrifying, suicidal passivity, on the other. For Douglass it was not enough simply to conceive of freedom as an abstract proposition and believe in the justice of it. The Nantucket convention made his freedom a concrete social reality, not just a theoretical principle or an act of faith. In Douglass's *Narrative* freedom is the social liberty *not* to have to live by a silent act of faith.

As Douglass refuses the tenets of American Protestantism, so he also breaks away from the characteristic topics, symbolism, and form of Christian narrative. Early narrators such as Gustavus Vasa and James Pennington wrote of their bondage and freedom wholly within the terms of the spiritual narrative. In Vasa's history, emancipation gives way to Christian conversion as the definitive event of his life. Pennington, although he granted much more importance to escape from slavery, similarly attributed definitive significance to his conversion. As he interpreted it, freedman's status merely led to the discovery that he was "a slave in another and a more serious sense," a "slave to Satan" in the spiritual bondage of sin. Douglass, however, refuses to let slavery acquire such a figurative or symbolic meaning in his narrative. He also repudiates the necessity for the transformation of being that is implied in the Protestant concept of spiritual regeneration. Although slavery can be a condition of the self, it does not define it. Pennington's narrative, whether intending to or not, asserts a discontinuity between the persona and the regenerate narrator, between slave and free author, as Pennington undergoes the transposition from literal to figurative slavery. Douglass insists, instead, that the author of his *Narrative* is the same human being who was enslaved, for he never speaks of a conversion in his life. If anything, he emphasizes the continuity between his free self and his enslaved self: "My feet have been so cracked with the frost, that the pen with which I am writing might be laid in the gashes." In this startling juxtaposition, the congruity of the fit reminds us of the literal meaning of slavery.

Douglass also refuses the formal constraints of Christian narrative, whose characteristic topic is the *via spiritualis,* the soul's journey to God. Although narrators such as Henry Bibb and William and Ellen Craft do not write about conversion, they still write indirectly within the constraints of the spiritual narrative. These authors structure their narratives as the story of the fugitive's flight, a materialistic inversion of the Christian soul's spiritual journey. As the Crafts' analogies to *The Pilgrim's Progress* suggest, these authors interpret their individual experience by literalizing the idealist metaphor. For example, William Craft describes his first sight of a free city, Philadelphia:

> The sight of those lights and that announcement made me feel almost as happy as Bunyan's Christian must have felt when he first caught sight of the cross. I, like him, felt that the straps that bound the heavy burden to my back began to pop, and the load to roll off. I also looked, and looked again, for it appeared very wonderful to me how the mere sight of our first city of refuge should have all at once made my hitherto sad and heavy heart become so light and happy. As the train speeded on, I rejoiced and thanked

God with all my heart and soul for his great kindness and tender mercy, in watching over us, and bringing us safely through.

Craft focuses entirely on his escape from literal bondage, but he still appeals to Bunyan's allegory to structure and interpret the significance of his experience.

Douglass makes no such appeal and his own fugitive's flight does not structure the contents of his *Narrative*. Rather, in Douglass's text we confront the silence of omission: "According to my resolution, on the third day of September, 1838, I left my chains, and succeeded in reaching New York without the slightest interruption of any kind. How I did so,—what means I adopted,—what direction I travelled, and by what mode of conveyance,—I must leave unexplained." Pointedly refusing to narrate his journey, Douglass calls our attention to the silence within which he made his escape, thus incorporating his refusal to speak as an intrinsic part of his *Narrative*. Strategically, he explains this omission is made on behalf of slaves who might yet escape, for secrecy of the means was essential. Rhetorically, this omission is equally purposeful: The audible silence of the ellipsis is the culmination of all the silent moments in the *Narrative*. It declares the invisible and unspoken presence of the enslaved; for rhetorically it represents the presence of the human being concealed in the language of slavery.

The words of the *Narrative* speak out of this silence. Douglass's eloquence has often been considered an astonishing accomplishment for an escaped slave, but something he must have learned on free soil. When we consider what Douglass himself says of language in slavery, another explanation seems more probable. The silence of the slave was not the silence of ignorance, much less of ineptitude, but the silence of a human being whose enslavement had forced on him an extraordinary knowledge of language use. The slave not only understood referentiality; he understood double referentiality. Moreover, he understood the critical importance of time, place, and audience in countless situations, and manifold possibilities of combining silence and sound. The linguistic virtuosity of the slave who survived slavery must have been impressive. The incentive to acquire a linguistic capability far beyond what was minimally necessary to labor in the fields was considerable, if only because the penalty for linguistic mistakes was incredibly high. The wrong word, nuance, or gesture at the wrong time could bring brutal punishment, even death. Considering what Douglass had *already* learned from his experience of language in slavery, to acquire some conventions of classical rhetoric must have seemed a small task.

Comprehending Slavery: Language and Personal History in the *Narrative*

John Sekora

> The author is therefore the more willing—nay, anxious, to lay alongside of such (pro-slavery) arguments the history of his own life and experiences as a slave, that those who read may know what are some of the characteristics of that highly favored institution, which is sought to be preserved and perpetuated.
>
> AUSTIN STEWARD

Because it is one of the most important books ever published in America, Frederick Douglass's *Narrative* of 1845 has justly received much attention. That attention has been increasing for a generation at a rate parallel to the growth of interest in autobiography as a literary genre, and the *Narrative* as autobiography has been the subject of several influential studies. Without denying the insights of such studies, I should like to suggest that in 1845 Douglass had no opportunity to write what (since the eighteenth century) we would call autobiography, that the achievement of the *Narrative* lies in another form.

Elsewhere I have argued the uniqueness of the antebellum slave narrative as an American literary form, a signal feature of which is the conditions under which it was printed. Briefly put, eighteenth-century narratives like those of Hammon, Gronniosaw, and Equiano were published only when they could be fitted to such familiar patterns as the captivity tale or the tale of religious conversion. In the

---

From *CLA Journal* 29, no. 2 (December 1985). © 1985 by the College Language Association.

abolitionist period when slavery was the central issue, once again printers and editors determined the overall shape of the narrative. Lundy, Garrison, Tappan, and Weld sought to expunge a vile institution, not support individualized Afro-American life stories. They had set the language of abolition—its vocabulary as well as social attitudes and philosophical presuppositions—in place by the early 1830s. Former slaves were wanted primarily as lecturers, later as authors, not for their personal identities as men and women, but for their value as eyewitnesses and victims. (It was significant to Douglass that his white associates tended to see slaves as passive victims.)

Against these conditions, one must place current conceptions of autobiography. Traditionalists and poststructuralists seem to agree that autobiography comes into being when recollection engages memory. Recollection engages people, things, events that at first appear fragmented and unrelated. As an essential part of its activity, recollection brings sequence and/or relation to the enormous diversity of individual experience; it emplots the stages of the subject's journey to selfhood. Meaning emerges when events are connected as parts of a coherent and comprehensive whole. Meaning, relation, and wholeness are but three facets of one characteristic: a narrative self that is more a literary creation than a literal, preexisting fact. The self of autobiography comes into being in the act of writing, not before. This said, the contrast with the antebellum narrative is apparent. From Hammon's *Narrative* of 1760 to Harriet Jacobs's *Incidents* in 1861, the explicit purpose of the slave narrative is far different from the creation of a self, and the overarching shape of that story—the facts to be included and the ordering of those facts—is mandated by persons other than the subject. Not black recollection, but white interrogation brings order to the narration. For eighteenth-century narratives the self that emerges is a preexisting form, deriving largely from evangelical Protestantism. For the abolitionist period, the self is a type of the antislavery witness. In each instance the meaning, relation, and wholeness of the story are given before the narrative opens; they are imposed rather than chosen—what Douglass in *My Bondage and My Freedom* (1855) calls "the facts which I felt almost everybody must know."

This approach would seem to resolve some persistent questions about the *Narrative*: why its structure is so similar to earlier (and later) abolitionist narratives, why Douglass subordinates so much of his emotional and intellectual life to the experience of slavery, why Garrison and Phillips are at such pains to make it appear a collective enterprise. At the same time it raises at least two others: if not autobiography, what kind of book is the *Narrative?* And how does it succeed so thoroughly?

White Americans, it would seem, have long attempted to cloak the raw experience of slavery—in the eighteenth century masking it in the language of triumphal Christianity, for most of the nineteenth century transmuting it into the language of abolition. For the years between 1870 and 1950, even this genteel transmutation was too raw, too threatening. Thus the slave narratives remained the most important and the most neglected body of early American writing. A consequence of that neglect is that we lack a distinctive term for a unique genre. For one example, most critics use the term *slave narrative* to refer to stories of oppression under slavery; yet before 1830 very few of the narratives concerned themselves with the injustices of the institution. Related to captivity tales, Franklinesque success stories, modern autobiographies, and other forms, the slave narrative is essentially different from all. It resembles other forms, but other forms do not resemble it.

In the absence of a distinguishing critical category, we must make do with that phrase used in authors' prefaces and advertisements as synonymous with abolitionist narrative—"personal history of slavery." Douglass was clearly aware in 1845 of the terms for such a history, for he had referred to them in his lectures earlier and wrote about them at length later. Before he became an antislavery agent, he had been questioned frequently concerning his life under slavery; once selected as an agent, he was coached concerning those aspects of slavery most likely to appeal to an ignorant or indifferent Northern audience. He had read the earlier separately published narratives and followed the shorter tales printed in the *Liberator* and other periodicals. He knew, he said, of the abolitionist emphasis upon facts, verifiable facts; upon instances of cruelty, repression, and punishment; upon the depth of Christianity in the owner's household; and so on. Overall, he knew that he was being woven into a network of clergymen, politicians, tradesmen, writers, editors, sponsors, and societies that was transatlantic in scope and resources. On any given day of lecturing, for example, he knew he would be introduced and followed by white speakers who would testify to his candor, character, and authenticity. And he knew he would conclude his address with an appeal to the audience to do as he had done—become absorbed in the abolitionist crusade.

Douglass was thus situated at the intersection of collectivizing forces. On both sides of the political divide, white people were busy defining and hence depersonalizing him. Apologists for slavery were doing their utmost to discredit him as a fraud; Garrison's agents were doing their best to publicize him as a representative fugitive slave. The issue over which they fought was not Douglass the lecturer or Douglass the author. Rather it was a narrower issue of their own defining. Douglass was important insofar as he embodied the experience of slavery. As author he was therefore caught in a genuine dilemma. He was indeed an individual human being with a particular story to tell, but if he were to discover personalizing words for his life, he must do so within the language of abolition. His success in resolving that dilemma, as arresting today as when it was first published, makes the *Narrative* the most comprehensive personal history of slavery in the language.

It embodies comprehension on several levels and in several successive stages, as intellectual apprehension of the many influences of slavery and narrative compassing its equally many forms. In the beginning Douglass as narrator comprehends the world that slaveholders have made, and Douglass as actor comprehends the power of language to transform that world. In his mature years he apprehends the eloquence of silence as well as the liberating power of words. Finally, as at once actor and narrator, he comprehends his own situation in the tradition of the slave narrative. Douglass highlights these levels with a series of gemlike sentences of Enlightenment irony and compression. Two are notable as preliminary illustrations of his modes of comprehension. Recalling his entrance into the Auld household in Baltimore twenty years before, he reports: "Little Thomas was told, there was his Freddy." In eight short words of indirect quotation, he signals both the effect of chattel slavery upon the Auld family and his intellectual apprehension of his place in the system. The Auld child is "Little Thomas"—an exalted owner of human property; Douglass is diminished in name as well as status—"*his* Freddy." In the kind of Enlightenment balance and compression sought by Hume and Johnson (but not surpassed by them), two adjectives modifying two nouns carry all the weight of significance: analysis enveloped by description. The Aulds address not Douglass but the child. Douglass is deployed as an object—in the sentence, the child's mind, the household, and the system—and so employs himself to convey his awareness of that situation. The burden of the narrative will be to reveal the necessary reversal of that situation. The

Aulds are too vacuous and vulnerable, Douglass too penetrating, for it to hold. In what Albert Stone has rightly called the key sentence of the *Narrative,* Douglass again unites balance, reversal, and narrative time to embody what an abolitionist narrative should be: "You have seen how a man was made a slave; you shall see how a slave was made a man." These two sentences suggest the depth of his enterprise. While operating within the abolitionist code, his adroit use of language would give his narrative a greater personal imprint, a wider historical compass, and a surer view of slavery than had ever been presented before.

That code prescribed that the opening portions of a narrative (as of a lecture) be heavily factual, containing if possible verifiable accounts of birth, parentage, and slaveholders. Douglass provided that—and much more. His opening paragraphs indicate a concern for accuracy designed to satisfy even the most hostile or scrupulous hunter of details. No one can do it better, he says in effect. He is then in position to portray the world of slaveholders and their minions, the world into which he was born. His plays upon the names of "Captain" Anthony, Mr. Severe, and Mr. Freeland are instances of his reduction of diverse personalities to their precise roles in an economic system they barely understand. With Austin Gore he provides a more elaborate description and a more powerful form of comprehension:

> Mr. Gore was proud, ambitious, and persevering. He was artful, cruel, and obdurate. He was just the man for such a place, and it was just the place for such a man.

With two trinities of adjectives and another sentence of finely wrought symmetry, the character of this baneful overseer is caught and reduced as if he were an overweening functionary in Molière or Ben Jonson. With Gore's employer, Colonel Lloyd, Douglass's irony is at full stretch. In Lloyd's callousness toward men and women and his sensitivity toward horses, he finds an apt sign of his owner's true worth. Because he is so utterly insecure with people, Lloyd's threat is shown to be hollow and his stature petty. When we learn that on his plantation only horses are treated with regard, we understand the social situation he has created and its underlying structure. He and his class are like the petty gods of Greek myth, absurd whichever way they turn.

In comprehending his own and their assistants, Douglass establishes his grasp of a type representing the most powerful families in the South. In a sense he has *defined* the type, as a dramatist does his primary actors. But he does not stop there, as earlier narrators had done. For his interweaving of interpretation with description has all along recognized the economic machinery of which Anthony and Lloyd are but small cogs. Slavery, he shows, wishes to control more than the labor and physical beings of slaves—even to their words, their very language. His first owner must be addressed as Captain Anthony: "a title which, I presume, he acquired by sailing a craft on the Chesapeake Bay." Like the many slaveholders who insisted upon being called "General" or "Colonel," Anthony demands that he be known by a self-conferred military title. Slaves alone could entitle masters, this artifice seems to say. Likewise, Colonel Lloyd rode out to outlying farms—where he wasn't known by sight—to question field slaves about how kindly their master was. For a candid answer, a slave would be sold South. Reporting these episodes, Douglass makes clear that what is being revealed is larger than his owner's self-deception. Slaveholders, by seeking to control slave language, sought to exact slave complicity in their own subjugation. Their self-conceptions required the right words, the correct words. With the proper words, a slave could keep his life intact. With the proper words, a slaveholder could keep his self-esteem intact. In each case, the owner compels the slave to authorize the owner's power. Slavery and the language of slavery are virtually coextensive.

In comprehending the equation of words and power, Douglass relates not only the workings of slavery as a system, but also the advent of his personal history within it. He describes his initial situation in chapter 2 as a well of ignorance, typified by his insensitivity to the words of work songs:

> I did not, when a slave, understand the deep meaning of those rude and apparently incoherent songs. I was myself within the circle; so that I neither saw nor heard as those without might see and hear. They told a tale of woe which was then altogether beyond my feeble comprehension.

This memory and the image of incomprehension spurred by it testify that for his life, as for the narrative we are reading, there will be no stop, no comforting return until his comprehension is complete. In one of the passages blending past and present at which he is so adept, he remarks:

The mere recurrence to those songs, even now, affects me; and while I am writing these lines, an expression of feeling has already found its way down my cheek. To those songs I trace my first glimmering conception of the dehumanizing character of slavery.

The next stage in his understanding of the language of slavery takes Douglass to Baltimore, the Auld household, and the forbidden seduction of reading. Auld's diatribe on the danger of language is well known as the impelling force for Douglass's climbing the ladder to literacy. Yet is is equally significant as a further sketch of the effects of slavery upon white people. Because they do not know what slavery is doing to them, the Aulds understand far less what it is doing to him. The exercise of petty power is for Mrs. Auld as corrupting as the possession of great wealth has been for Colonel Lloyd. (It is also possible to see in her decline features of those Garrisonians who turned on Douglass.) It is through them that Douglass gains his penultimate lessons of the perversions of slavery.

In most abolitionist narratives the quest for freedom through literacy would conclude here, the story redirected toward plans for escape. With Douglass, however, simple literacy is merely the ground upon which a complex psychological drama will be played. Although he has learned much from earlier narratives, he will not provide exactly the same kind of straight-line narrative found in, say, Moses Roper, James Curry, Lunceford Lane, or Moses Grandy. His war with slavery through language consists not of a single battle with clear-cut victory on either side. Rather it is a sustained series of costly skirmishes, with losses following hard upon gains. As Auld had predicted, Douglass at twelve years of age is beset by discontent. His fall is occasioned by his hunger for language, for while his readings "relieved me of one difficulty, they brought on another even more painful than the one of which I was relieved." The more he learns of slavery, the farther freedom seems to recede. The condition is temporary since he refuses to be satisfied. The discontent brought on by language will be relieved by language: in this instance by a single word—*abolition*—and its resonance. The pain that is aggravated by language is also palliated by language. As he came to apprehend the meaning of abolition, he records, "The light broke in upon me by degrees."

The predicament recurs at a higher level when he is broken by

Covey's demands of incessant labor: "I was broken in body, soul, and spirit. . . . [T]he dark night of slavery closed in upon me." The language of abolition he has been learning possessed the power to inspire longing and to instill despair when that longing is thwarted. And once again a call in words evokes a powerful response, in the apostrophe to the ships on the Chesapeake so well analyzed by Stone. His predicament is resolved in a form of blues sermon that raises all doubts and answers all: "There is a better day coming." Douglass's career has been an ascent toward freedom through literacy. His comprehension of the language—first of slavery, then of abolition—has been the ladder of his climb. Structurally, he himself marks his rise to the top by his battle with Covey and the pivotal sentence, "You have seen how a man was made a slave; you shall see how a slave was made a man." Thematically it is the "protections" he writes for himself and others in 1835 that signal his position. With the protection he can *write* his way North, the ultimate verification of his victory over slavery and a final proof of his comprehension of language.

Although the contest with Covey has made chapter 10 the most famous portion of the *Narrative,* it is the final chapter that most reveals its distinction as a personal history. In half the length of the preceding section, chapter 11 accomplishes three very large tasks of comprehension. First, he exercises an eloquence of silence fully as powerful as his brilliance of language. When he forgoes an account of his escape, he relinquishes an element of the narrative that had made it one of the most popular literary forms in America in the 1840s. For example, in Moses Roper before him, William Wells Brown, Henry Box Brown, the Crafts, and John Thompson after him, the escape is an exciting adventure story in itself, uniting ingenuity, suspense, courage, and endurance. It is, in short, precisely the kind of story Northeastern audiences would pay to hear and read. Douglass's decision to withhold that part of his story is an assertion of personal control within a mandated form. Only he can write this section, not Garrison or Phillips; only he knows what is being withheld. Only he can decide the proper time for its release. At the moment of writing he is painfully aware of the short distance (political as well as temporal) that separates past from present. Hence his silence is evoked by a communal regard for fellow slaves still seeking means of escape: he must "not run the hazard of closing the slightest avenue by which a brother slave might clear himself of the chains and fetters of slavery."

Second, what he does choose to include equally bears his personal stamp, the language of a free man. The two sentences in which he dates his escape and destination are models of laconic understatement, conspicuous in their restraint. By this point he has comprehended the art of the oxymoron, as he provides readers with poised anxiety and loud softness—eloquent silence. Also conspicuous is his inclusion of a second document (the first being the protections), the only one not of his own composition. The certificate of marriage to Anna Murray, subscribed by James W. C. Pennington, appears to be Douglass's proof in language of a new existence. Socially, legally, sexually, religiously, he is indeed a man—in the eyes of most Americans, for the first time. It was to this need for documentation that Pennington returned in his narrative, *The Fugitive Blacksmith,* in 1849. In his preface, he wrote:

> Whatever may be the ill or favored condition of the slave in the matter of mere personal treatment, it is the chattel relation that robs him of his manhood, and transfers his ownership in himself to another. . . . It is this that throws his family history into utter confusion, and leaves him without a single record to which he may appeal in vindication of his character, or honor. And has a man no sense of honor because he was born a slave? Has he no need of character?

Douglass has ensured that his new family will be recorded, will from its inception possess a sense of honor.

In his final narrative gesture, Douglass establishes that he comprehends the tradition of the slave story and attempts to subvert a portion of that tradition. By closing with his address to the Nantucket convention in 1841, Douglass—as Stone and Stepto have cogently argued—brings the narrative full circle, to the opening sentence of Garrison's preface. Garrisonians, he explains in *My Bondage and My Freedom,* often sought to limit his scope: "Give us the facts . . . we will take care of the philosophy." Here he makes no mention of Garrison and reverses a persistent abolitionist tactic. Garrison and his associates often spoke as if former slaves were minor characters in *their* great antislavery story. Douglass deftly ensures that Phillips and Garrison will, in this narrative, be minor characters in *his* story. He authenticates them.

It is a bold gesture. For on a philosophical level, one might say that the slave narrative as a form is defined paradoxically by a suppression of the personal voice of the slave. Most sponsors regarded the slave by stipulation as primitive and then proceeded to use the narrative to address other white people. Many sponsors condescendingly saw the narratives as essentially a political form for their own use and said that fugitive slaves had no stories until the abolitionists gave them one. Douglass by 1845 is certainly aware of the complex of attitudes surrounding him: "The truth was, I felt myself a slave, and the idea of speaking to white people weighed me down." Humility and restraint are poised in this final paragraph, for it gains dignity by using Enlightenment language when explosive effusion seems called for. Like his audience in Nantucket, his readers acknowledge the effort, the discipline of his control. Slavery is far worse than anything he can say about it. The tension created between the cruelties which he recounts and his manner of recounting them he will use communally, not to win applause, but to go on working. The surplus of tension will be spent in the future and in language; "engaged in pleading the cause of" his "brethren." Whatever his sponsors intend, he will not be distracted. His tension will be active, always on the move, always renewing and being renewed.

The *Narrative*, I would contend, is the first comprehensive, personal history of American slavery. Autobiography would come a decade later, in *My Bondage and My Freedom.* If many readers prefer the earlier volume, the reasons are not far to search. The *Narrative* is as tightly written as a sonnet, the work of years in the pulpit and on the lecture circuit. It comprehends all major aspects of slavery as Douglass knew it in a narrative that is as dramatically compassing as any first-person novel. It is at the same time a personal history of the struggle with and for language—against words that repress, for words that liberate. It is for author and reader alike a personalizing account of a system that would depersonalize everyone. It is the retelling of the most important Christian story, the Crucifixion, in the midst of the most important American civil crisis, the battle over slavery.

In *The Fugitive Blacksmith* Pennington asked if a slave had no need of character. He answered the question in the following way: "Suppose insult, reproach, or slander, should render it necessary for him to appeal to the history of his family in vindication of his character, where does he find that history? He goes to his native state, to his native country, to his native town; but nowhere does he find any record of himself *as a man*." It is an acute question, one he is eager to raise, I believe, because of Douglass's example. Douglass renewed the conservative form of the slave narrative at a critical time. He gave record of himself as an antislavery man. And the magnitude of that achievement is difficult to overestimate. For in moral terms the slave narrative and its postbellum heirs are the only history of American slavery we have. Outside the narrative, slavery was a wordless, nameless, timeless time. It was time without history and time without imminence. Slaveholders sought to reduce existence to the duration of the psychological present and to mandate their records as the only reliable texts. Whatever the restrictions placed upon them, Douglass and the other narrators changed that forever. To recall one's personal history is to *renew* it. The *Narrative* is both instrument and inscription of that renewal.

The Performance of the *Narrative*

William L. Andrews

When he wrote his *Narrative* in 1845, Frederick Douglass was not indifferent to the ideological consensus that the narratives of black "men of propriety" like Grandy, Lane, and Henson tried to establish with the white middle-class reader of the North. In his own unprecedented way, Douglass participated in and appealed to this consensus by fashioning his autobiography into a kind of American jeremiad. This genre differs from that which Wilson J. Moses has labeled the black jeremiad in one crucial sense: while the latter was preoccupied with America's impending doom because of its racial injustices, the American jeremiad foretold America's future hopefully, sustained by the conviction of the nation's divinely appointed mission. The practitioners of both literary traditions tended to see themselves as outcasts, prophets crying in the wilderness of their own alienation from prevailing error and perversity. While the white Jeremiahs decry America's deviation from its original sacred mission in the New World, they usually celebrate the national dream in the process of lamenting its decline. The American jeremiad affirms and sustains a middle-class consensus about America by both excoriating lapses from it and rhetorically coopting potential challenges (such as those offered by Frederick Douglass) to it.

With Herman Melville and Henry David Thoreau, contemporary Jeremiahs who also addressed the national sin of slavery, Doug-

---

From *To Tell a Free Story*. © 1986 by the Board of Trustees of the University of Illinois Press. University of Illinois Press, 1986.

lass confronted America with profoundly polarized emotions that produced in him a classic case of Du Boisian double consciousness. As a fugitive slave orator in the early 1840s, he denounced the institutionalized racism that pervaded America and perverted its much-heralded blend of liberty, democracy, and Christianity. Following the Garrisonian line, his speeches poured contempt on the Constitution of the United States as a compact with slavery and condemned northern as well as southern Christians for being the slave's tyrants, "our enslavers." The *Narrative,* however, goes to no such political or religious extremes. In that book, Douglass deploys the rhetoric of the jeremiad to distinguish between true and false Americanism and Christianity. He celebrates the national dream by concluding his story with a contrast between the thriving seacoast town of New Bedford, Massachusetts—where he was "surrounded with the strongest proofs of wealth"—and the run-down Eastern Shore of Maryland, where "dilapidated houses, with poverty-stricken inmates," "half-naked children," and "barefooted women" testified to an unprogressive polity. Appended to Douglass's story is an apparent apology for his narrative's "tone and manner, respecting religion," but this quickly gives way to a final jeremiad against the pharisaical "hypocrites" of "the *slaveholding religion* of this land." "I love the pure, peaceable, and impartial Christianity of Christ," Douglass proclaimed. All the more reason, therefore, for him to appropriate the language of Jeremiah 6:29 for his ultimate warning to corrupters of the faith: " 'Shall I not visit for these things? saith the Lord. Shall not my soul be avenged on such a nation as this?' " The *Narrative* builds a convincing case for Douglass's literary calling and his ultimate self-appointment as America's black Jeremiah.

Douglass's account of his rise from slavery to freedom fulfills certain features of the jeremiad's cultural myth of America. The *Narrative* dramatizes a "ritual of socialization" that Sacvan Bercovitch finds often in late eighteenth- and early nineteenth-century jeremiads: the rebellion of a fractious individual against instituted authority is translated into a heroic act of self-reliance, a reenactment of the national myth of regeneration and progress through revolution. The great rhetorical task of the jeremiad is to divest self-determinative individualism of its threatening associations with anarchy and antinomianism, the excesses of the unbridled self. In America the jeremiad made much of the distinction between rebellion and revolution. The rebel disobeys out of self-interest and defiance of the good of the community and the laws of Providence. His act parallels Lucifer's primal act of disobedience, which produced only discord and a (temporary) thwarting of the divine plan. The revolutionary, on the other hand, promotes in the secular sphere the same sort of upward spiral toward perfection that God demanded of each individual soul in its private progress toward redemption. The American jeremiad obviated the distinction between secular and sacred revolution in order to endow the former with the sanction of the latter, the better to authorize the national myth of the American Revolution. America was a truly revolutionary society in the sense and to the extent that its people—that is to say, those who had been accorded the status of personhood in the Constitution—remained faithful to God's plan for the progressive conversion of their land into a new order. Americans were therefore called to be revolutionaries, but revolutionaries in the service of an evolving divine order within which Americans could achieve corporate self-realization as God's chosen people.

As several critics have noted, Douglass's *Narrative* seems to have been consciously drawn up along structural and metaphorical lines familiar to readers of spiritual autobiographies. The young Frederick is initiated into a knowledge of the depravity of man when he witnesses the hideous flogging of his aunt Hester. "It was the blood-stained gate, the entrance to the hell of slavery, through which I was about to pass." Though seemingly damned to this southern hell, the eight-year-old boy is delivered by "a special interposition of divine Providence in my favor" from the plantation of Edward Lloyd in Talbot County to the Baltimore home of Hugh Auld. From that time forward, the boy is convinced that freedom—"this living word of faith and spirit of hope"—would be his someday. The thought "remained like ministering angels to cheer me through the gloom." Douglass's faith in this intuitively felt heavenly promise of liberation undergoes a series of trials in his boyhood and early teens, when he is first led out of his "mental darkness" by Sophia Auld, who teaches him his letters, and then is thrust back into "the horrible pit" of enforced ignorance by her husband, who fears a mentally enlightened slave.

The middle chapters of the *Narrative* recount the slave youth's growing temptations to despair of deliverance from bondage. Returned to the rural region where he was born, Douglass discovers the hypocrisy of Christian slaveholders whose pretentions to piety mask their cruelty and licentiousness. He reaches a dark night of the soul in 1833, when the harsh regime of Edward Covey, "the snake," breaks him "in body, soul, and spirit." And yet, he undergoes "a glorious resurrection, from the tomb of slavery, to the heaven of freedom" by violently resisting Covey's attempt to apprehend him, one August morning, for another infraction of the rules. Thus, "resurrected," the sixteen-year-old youth, "a slave in form" but no longer "a slave in fact," begins to put his revived faith in freedom and his "self-confidence" into practice. Hired out in 1834 to William Freeland, he starts a "Sabbath school" in which to teach slaves how to read the Bible and "to imbue their minds with thoughts of freedom." "My tendency was upward," states Douglass, firmly committed to following the road to freedom analogized in the *Narrative* in imagery reminiscent of *Pilgrim's Progress*. The first escape attempt, appropriately timed for Easter, is foiled, but his second, in September 1838, succeeds.

The last pages of the *Narrative* describe the new freeman's call to witness for the gospel of freedom that had preserved, regenerated, and pointed him northward. A subscription to the *Liberator* sets his "soul" on fire for "the cause" of abolitionism. At an antislavery meeting in Nantucket in August 1841, "I felt strongly moved to speak," but Douglass is restrained by a sense of unworthiness before white people. Still the promptings of the spirit cannot be resisted, even though it is "a severe cross" for the new convert to take up. "I spoke but a few moments, when I felt a degree of freedom, and said what I desired with considerable ease." This liberation of the tongues climaxes the life-long quest of Frederick Douglass toward his divinely appointed destiny in the antislavery ministry. The special plan of Providence is now fully revealed at the end of the *Narrative*. Frederick Douglass is a chosen man as well as a freeman. His trials of the spirit have been a test and a preparation for his ultimate mission as a black Jeremiah to a corrupt white Israel. This autobiography, as Robert G. O'Meally has emphasized, is a text meant to be preached.

Like all American jeremiads, the *Narrative* is a political sermon. Douglass's self-realization as a freeman and a chosen man takes place via a process of outward and sometimes violent revolution as well as inner evolution of consciousness. The strategy of Douglass's jeremiad is to depict this revolution as a "process of Americanization," to use once again a key phrase in Bercovitch's analysis of the genre. As Bercovitch notes, the jeremiad was responsible for rationalizing and channeling the revolutionary individualistic impulse in America so as to reconcile it with the myth of America's corporate destiny as a chosen people. This meant distinguishing firmly between the truly American revolutionary individualism and the rebellious, un-American individualism of the alien and seditious Indian, Negro, or feminist. Those marked by racial heritage as other had to *prove* that they were of "the people," the American chosen, by demonstrating in their own lives the rituals of Americanization that had converted them from nonpersons, as it were, into members of the middle-class majority. "Blacks and Indians . . . could learn to be True Americans, when in the fullness of time they would adopt the tenets of black and red capitalism." Along with Lane and Henson, Douglass pledges allegiance to the economic tenets of the republic in his autobiography, entitled, appropriately enough, the narrative of *"An American Slave."* Douglass goes beyond either Lane or Henson, however, in using his orthodoxy to justify his revolution against slavery and its perverse, un-American profit motive.

John Seelye has pointed out some of the affinities between Douglass's *Narrative* and the cultural myth of America as dramatized in Franklin's memoir. Douglass is "Ben Franklin's specific shade," argues Seelye, though the ex-slave's story is "not a record of essays to do good but attempts to be bad, Douglass like Milton's Satan inventing virtue from an evil necessity." It is no small part of Douglass's rhetorical art, however, to translate his badness into revolutionary necessity of a kind his white reader could identify with. Douglass's life, like Franklin's, describes a rising arc from country to city; then it follows a downward curve of expectations when the town slave is returned to the plantation; but it revolves upward once more, after the fight with Covey, carrying the black youth from Talbot County to Baltimore and finally to New York and New Bedford. When given the opportunity for self-improvement in the city, young Fred is every bit as enterprising as Father Ben. As a boy he overcomes adversity to learn reading and writing on his own. Discovery of an eloquency handbook entitled *The Columbian Orator* foreshadows the day when he will become one. But first he must become a man. The battle with Covey halts his regression into the slave's "beast-like stupor" and "revived within me a sense of my own manhood." "Bold defiance" replaces "cowardice"; significantly, the spheres of this defiance are intellectual and economic. As a Sabbath school teacher at Freeland's, Douglass uses as his text the Bible, and his aim is consistent with America's middle-class civil religion. He encourages his slave pupils to behave "like intellectual, moral, and accountable beings" rather than "spending the Sabbath in wrestling, boxing, and drinking whisky." Douglass domesticates a greater gesture of defiance, his first escape attempt, by analogizing it to the hallowed decision of Patrick Henry (ironically, a slaveholder) for liberty or death. "We did more than Patrick Henry," the fugitive advances of himself and his fellow runaways, because their liberty was more dubious and their deaths more certain if they failed. By implication, then, these dauntless blacks were more heroically American in their struggle for independence than was one of the most prominent delegates of the convention of 1776.

After Douglass's return to Baltimore in the spring of 1835 to be hired out in various employments, the *Narrative* concentrates increasingly on the economic humiliations of an upwardly aspiring "slave in form but not in fact." While apprenticed as a ship's carpenter, the slave is attacked and beaten by four white workers who felt it "degrading to them to work with me." Manfully, Douglass returns the blows in kind despite the adverse odds. Taught the calking trade, he "was able to command the highest wages given to the most experienced calkers," $1.50 a day. The money "was rightfully my own," Douglass argues. "I contracted for it; I earned it; it was paid to me." Yet every week Hugh Auld took it. "And why? Not because he earned it,—not because he had any hand in earning it,—not because I owed it to him,—nor because he possessed the slightest shadow of a right to it; but solely because he had the power to compel me to give it up." With Grandy, Lane, and Henson, Douglass appeals to his reader's respect for contract and resentment of arbitrary power as a way of preparing his case for the final break with slavery. The right at issue here is pragmatic and economic, not abstract or romantic. Douglass analogizes Auld to a "grim-visaged pirate" and a "robber"—an outlaw, in other words—to banish him from a consubstantial relationship with the northern reader. Meanwhile, Douglass qualifies himself for acceptance as an economic revolutionary in the best American tradition. He works his way up the economic ladder in the South from country slave to city apprentice to the quasi-free status of one who "hired his time" (the situation of Grandy, Lane, and Henson before they extricated themselves from slavery).

Hiring his time required Douglass to meet all his living expenses out of the income that he could make for himself, while still paying his master a fixed return of $3 per week. This was "a hard bargain," but still "a step toward freedom to be allowed to bear the responsibilities of a freeman." Here again, Douglass stresses how he qualified himself, step by step, for freedom and its "responsibilities" as well as its "rights." "I bent myself to the work of making money," adds Douglass, by way of proving his dedication to the quintessential responsibilities of an American free man. "I was ready to work at night as well as day, and by the most untiring perseverance and industry, I made enough to meet my expenses, and lay up a little money every week." The savings were used, presumably, to help Douglass in his flight to freedom, for less than a month after Auld halted the slave's hiring out (fearing that too much freedom would go to the black man's head), Douglass retaliated by taking the ultimate "step toward freedom."

One of the most unconventional features of the *Narrative* was Douglass's refusal to end his story with the stock-in-trade climax of the slave narrative. Watching the panting fugitive seize his freedom just ahead of snapping bloodhounds and clutching slavecatchers left white readers with a vicarious sense of the thrill of the chase as well as the relief of the successful escape. In the slave narrative a generation of readers found a factual parallel to the capture-flight-and-pursuit plots of their favorite romances by James Fenimore Cooper, William Gilmore Simms, and Robert Montgomery Bird. Yet Douglass left only a hiatus in his story where the customary climax should have been, insisting, quite plausibly, that to recount his mode of escape would alert slaveholders to it and thus close it to others. The conclusion he chose for his *Narrative* indicates that in his mind the high point of a fugitive slave's career was not his arrival in the free states but his assumption of a new identity as a free man and his integration into the American mainstream.

Douglass notes graphically the initial terrors of the isolated fugitive in a strange and often hostile land, but his emphasis is on how quickly and happily he assimilated. He marries within two weeks of his arrival in New York. He and his wife Anna move immediately to New Bedford, where the morning after his arrival he receives from his Negro host a new name to denote his new identity in freedom. Two days later, he takes his first job stowing a sloop with a load of oil. "It was the first work, the reward of which was to be entirely my own." "It was to me the starting-point of a new existence."

Everything falls into place for Douglass in New Bedford, where the American dream of "a new existence" is always possible for every man, black or white. New Bedford fulfills the ex-slave's socioeconomic quest; here every man pursues his work "with a sober, yet cheerful earnestness, which betokened the deep interest which he felt in what he was doing, as well as a sense of his own dignity as a man." Most marvelous of all, the black population of this paragon of industrial capitalism lives in "finer houses" and enjoys "more of the comforts of life, than the average of slaveholders in Maryland." True, Douglass admits, "prejudice against color" along the docks of New Bedford kept him from resuming his former trade as a calker. But a note to the text removes even this blemish from the image of the town as the epitome of progress and justice: "I am told that colored persons can now get employment at calking in New Bedford—a result of anti-slavery effort." Perhaps this is the reason for the mild manner and the absence of irony or bitterness with which Douglass brings up this lone instance of racism in the North. The refusal of New Bedford's calkers to work with him moves the narrator to none of the moral outrage that accompanies his recall of the same kind of treatment that he received from Baltimore's calkers. Now Douglass is more thick-skinned and matter-of-fact; his narrative business is not to complain about the barriers to his progress but to show how he, like his adopted city, overcame them. Now is the time for understatement: "Finding my trade of no immediate benefit, I threw off my calking habiliments, and prepared myself to do any kind of work I could get to do."

For the next three years, Douglass had to support his family via whatever manual labor jobs he could find, including sawing wood, shoveling coal, sweeping chimneys, and rolling casks in an oil refinery. Yet the *Narrative* stresses only the bright side of this experience—Douglass's American ingenuity and industry—not the ugly side—New Bedford's economic repression of a trained black tradesman. Only in 1881, in his *Life and Times,* when Douglass no longer had the same rhetorical stake in a dramatic contrast between North and South, would he call the whole humiliating episode "the test of the real civilization of the community" of New Bedford, which the town plainly failed. In 1845 New Bedford had to serve as Douglass's standard of "real civilization," of true Americanism, so that he as a jeremiad writer could have something by which to measure the South's fall from national grace.

Like earlier popular literary genres from which Afro-American autobiography sought authentication and other rhetorical advantages, the American jeremiad provided a structure for Douglass's vision of America that was both empowering and limiting at the same time. The jeremiad gave the ex-slave literary license to excoriate the South pretty much as he pleased so long as the ideals and values by which he judged that region's transgressions remained American. Thus while bitterly evoking the nightmare of slavery, Douglass's example invoked just as reverently the dream of America as a land of freedom and opportunity. In a letter to Douglass several weeks before the *Narrative* was published, Wendell Phillips, one of the most forthright abolitionist critics of racism in the North as well as slavery in the South, urged the autobiographer to include a comparison of the status of blacks in both sections of the country. "Tell us whether, after all, the half-free colored man of Massachusetts is worse off than the pampered slave of the rice swamps!" Phillips requested, with his usual penchant for irony. In Douglass's jeremiad, however, such a topic was not tellable. In the spiritual autobiography and the success story, of which Douglass's *Narrative* is an amalgam, doubts about the achievement or significance of salvation and success are clear evidence that they have not been attained. Douglass's story, by contrast, is determined to declare New Bedford as more than one slave's attainable secular salvation in America. Such a declarative act brought into being New Bedford as Douglass needed it to be—a symbol of his belief in America as a free, prosperous, and progressive social order that thrived without caste distinctions or the exploitation of labor. For the sake of this symbol in his vision of America, Douglass could make his own exploitation in the New Bedford labor market seem like a useful lesson in the school of hard knocks, the sort of adversity that self-made men generally glory in. For the symbol's sake, Douglass would censor himself and say nothing of more humiliating Jim Crow experiences that he had been subjected to in the North, although he had been recounting such incidents from the abolitionist platform for the past three years.

The American jeremiad structured Douglass into a fixed bipolar set of alternatives with which to define the experience and aspirations of "an American slave." As a revealed truth represented symbolically, America in the jeremiad could be understood only in terms of "alternatives generated by the symbol itself." That which was not American was conceived of as an absence, un-Americanism, false Americanism. America was constantly being analyzed and measured against its opposite, which was only the negative function of the interpretive possibilities of the symbol. To get outside this self-enclosed heuristic dualism, one had to liberate oneself from the symbol of America as a self-valorizing plenitude and from the binary oppositions that maintained the symbol within a field of meanings of its own making. Henry Louis Gates, Jr., has argued convincingly that in the first chapter of the *Narrative,* the binary oppositions that inform and enforce the culture of the slavocracy are "mediated" by the narrator so as to "reverse the relations of the opposition" and reveal that "the oppositions, all along, were only arbitrary, not fixed." For instance, as both the son and slave of his father-master, the mulatto Douglass deconstructs the fundamental opposition between white people and black animals on which much of the rationale for slavery was based. That separation between white and black cannot hold because it is culturally, not naturally, determined. By the time we finish the last chapter of the *Narrative,* however, it becomes evident that Douglass is not bent on the same kind of critique of the binary oppositions that govern and validate the symbol of America. The *Narrative* turns, structurally and thematically, on such dualities as southern slavery versus northern freedom, "slaveholding religion" versus "Christianity proper," Baltimore versus New Bedford, compulsion versus contract, stagnation versus progress, deprivation versus wealth, violence versus order, community versus caste system. And very little mediation takes place between these fixed, shall we say "black-and-white," antitheses. Indeed, Douglass suggests that the gap between these poles of true and false Americanism is growing wider, as New Bedford's progress against racial discrimination seems to testify.

Thus as an American jeremiad, Douglass's *Narrative* deconstructs binary oppositions that uphold slavery in the South while reconstructing the pattern of his life around other sets of oppositions whose support of the myth of America he might as readily have questioned, too. In 1845, however, Douglass was still exploring the heuristic and rhetorical possibilities of binary oppositions as a means of establishing his own identity relative to America, South and North. It is through his own experiments with rhetoric that we see Douglass's particular brand of "opposing self" at work. As a jeremiadic autobiographer, he has more than his own story to tell. He must preach in such a way as to discredit the false oppositions and hierarchies of value that have arisen as a consequence of slavery's perversions of the true oppositions between good and evil, the natural and the unnatural. This is the major reason for Douglass's self-conscious introduction of traditional tropes of rhetoric into the slave narrative.

Douglass followed the figural convention of earlier tropological black autobiographers when he appropriated from Christian theology the metaphors of spiritual *revolutio* that let him convert his violent resistance to Covey into a "glorious resurrection, from the tomb of slavery." But analogical argument could not help the ex-slave expose the stark inconsistencies between southern practice and the American promise or the inversions of nature and value that slavery forced. Hence his regular use of paradox, hyperbole, chiasmus, and other varieties of antithetical clausal constructions. Captain Thomas Auld "was a slaveholder without the ability to hold slaves." Slaves "sing the most pathetic sentiment in the most rapturous tone, and the most rapturous sentiment in the most pathetic tone." The overseer Austin Gore "dealt sparingly with his words, and bountifully with his whip." Once "the fatal poison of irresponsible power" was put in Sophia Auld's hands, "that cheerful eye, under the influence of slavery, soon became red with rage; that voice, made all of sweet accord, changed to one of harsh and horrid discord; and that angelic face gave place to that of a demon." Covey was such a harsh taskmaster that "the longest days were too short for him, and the shortest nights too long for him."

Had Douglass confined himself to these tropes for the reinforcement of his criticism of an inverted and perverted culture, he might not have troubled a man like Ephraim Peabody very much. But Douglass's tropological action is much more diverse and includes theatrical and other "playful" effects that are only loosely bound to the service of fact and theme. Repetition is Douglass's favorite rhetorical technique for experimentation. In the first chapter he uses it within a structure of antitheses to impress upon his reader the immitigable, metronomic rhythm of a whipping.

> I have often been awakened at the dawn of day by the most heart-rending shrieks of an own aunt of mine, whom he [Captain Anthony, young Fred's first master] used to tie up to a joist, and whip upon her naked back till she was literally covered with blood. No words, no tears, no pray-

> ers, from his gory victim, seemed to move his iron heart from its bloody purpose. The louder she screamed, the harder he whipped; and where the blood ran fastest, there he whipped longest. He would whip her to make her scream, and whip her to make her hush; and not until overcome by fatigue, would he cease to swing the blood-clotted cowskin.

Here Douglass repeatedly "whips" his reader with a word in order to induce in that reader a response to a literary figure that will be in some slight sense analogous to his aunt's experience of being beaten. This is rhetoric used for a traditional purpose in the slave narrative: to put the reader in the place of someone with whom he or she ought to sympathize. Whipping is a fact of slavery of which the white reader must have empathic as well as objective knowledge.

Consider, by contrast, Douglass's later repetitious use of the same word in characterizing the Reverend Rigby Hopkins of Talbot County.

> Mr. Hopkins could always find some excuse for whipping a slave. It would astonish one, unaccustomed to a slaveholding life, to see with what wonderful ease a slaveholder can find things, of which to make occasion to whip a slave. A mere look, word, or motion,—a mistake, accident, or want of power,—are all matters for which a slave may be whipped at any time. Does a slave look satisfied? It is said, he has the devil in him, and it must be whipped out. Does he speak loudly when spoken to by his master? Then he is getting highminded, and should be taken down a buttonhole lower. Does he forget to pull off his hat at the approach of a white person? Then he is wanting in reverence, and should be whipped for it. Does he ever venture to vindicate his conduct, when censured for it? Then he is guilty of impudence,—one of the greatest crimes of which a slave can be guilty. Does he ever venture to suggest a different mode of doing things from that pointed out by his master? He is indeed presumptuous, and getting above himself; and nothing less than a flogging will do for him. Does he, while ploughing, break a plough,—or, while hoeing, break a hoe? It is owing to his carelessness, and for it a slave must always be whipped.

In this passage Douglass harps on the word "whip" in order to sting Hopkins and his ilk with their own cruel consistency. The narrator's mockery depends on several levels of repetition: linguistic, syntactic (the "Does he . . . ? Then he. . . ." construction), and thematic (everything the slave does supposedly signifies the same thing and must receive the same response). Douglass's genius lies in the stage management of these effects so that they will do their maximum moral damage to the slaveholders while accruing literary benefits to the slave narrator as satirist.

Douglass begins with a come-on, the promise of something "astonishing," and then states his sweeping thesis—that virtually anything the slave does can be used as an excuse for a beating. Instead of listing the possible occasions for the whip in a plain and factual manner, however, Douglass sets up a dramatic structure in which he impersonates both sides of an imaginary dialogue, or question-and-answer session. As questioner Douglass plays the role of an objective observer of phenomena who inquires innocently about their meaning, somewhat as the curious white reader of the North might do. As respondent Douglass, now the southern white authority for the northern innocent, answers in such a repetitive and compulsive way that his explanatory authority is swept away by his reductive cant. Douglass surveys the vernacular and official modes of explaining slave behavior, from "he has the devil in him" to "he is indeed presumptuous," but the more inflated the terms become, the more the pretense of it all is exposed by its automatic translation into the same gross and brutal consequence, the whipping. The outcome of the satire is reductio ad absurdum. The slaveholder's reductive reading of his slave is ironically reversed to reduce *him* to the level of a comic grotesque.

In both the Anthony and the Hopkins passages, Douglass takes a "brute fact" (linguistically and morally) like the whipping of slaves and experiments with rhetorical contexts in which to turn that fact to his own expressive account. Simple assertions about the method, incidence, or justifications of whipping are insufficient to this slave narrator's purposes. Nor is he satisfied to couch the fact of whipping in a standard expressive speech act that will convey his own psychological response in the hope of prompting his reader to similar feelings. Instead, Douglass makes a crucial decision: to present the fact of whipping to the reader in two deliberately stylized, plainly rhetorical, recognizably artificial contexts. There is nothing masked about this presentation. On the contrary, Douglass's choice of repetition as his chief rhetorical effect in both passages leaves his mark unmistakably on the text in bold strokes that constitute his stylistic signature.

In passages like these, Douglass calls attention to himself as an unabashed artificer, a maker of forms and effects that recontextualize brute facts according to requirements of self. The freeman requires the freedom to demonstrate the potency of his own inventiveness and the sheer potentiality of language itself for rhetorical manipulation. As Douglass develops his style, he gains literary mastery over the brute facts of the slave past. As he deliberately exhibits that style, he repossesses autobiography as a self-expressive, not simply a fact-assertive, act. This does not mean of course that the style of autobiography provides the key to Douglass's essential self. What is expressed through the style of the Anthony and Hopkins passages is Douglass's performing self, plainly and exuberantly engaged in performing rhetorical operations on brute facts and consciously aware of itself in the process as a player of roles, a maker of effects, and a manipulator of readers.

William Lloyd Garrison's preface to the *Narrative* promises its reader that Douglass had been "essentially true in all [his] statements; that nothing has been set down in malice, nothing exaggerated, nothing drawn from the imagination." In Garrison's sense of the word, the imagination was the wellspring of fabrications and exaggerations, and hence not to be acknowledged as part of a slave narrator's intellectual resources. Today, questions about whether Douglass imagined or exaggerated matters in the *Narrative* continue to be researched by historians. What is more to our purpose is Douglass's employment of his imagination, his ability to portray images in language and thereby evoke sensations in his reader, as a means of influencing his reader's perceptions of and response to him and his world. The Anthony passage exemplifies in the *Narrative* what Philip Wheelwright has termed the "confrontive imagination." The aim of Douglass's style in that instance is to confront the reader with particular details of a specific whipping, not the general idea of whipping. Through an imaginative use of rhetoric, Douglass attempts "to intensify the immediate experience itself," to bring it home to the reader, as it were. By contrast, in the Hopkins quotation, Douglass is engaged in the "imaginative distancing" of his reader from the subject at hand. Douglass's style is disassociative in that his humor reduces and removes Hopkins from the normal relationship of respect that he would have with the white reader. Douglass's satiric humor carries the reader away from its object while the tragic seriousness of the confrontative style draws the reader to its object.

Douglass's rhetoric, aided by these two modes of imagining, is concerned, therefore, with moving its reader not just emotionally but also spatially with respect to the text. Traditionally, I have argued, black autobiography tried to move its white reader in one direction, from an alien to a consubstantial relationship with the text and the black self presumably represented by the text. Douglass's rhetoric ran the risk, according to Ephraim Peabody, of reversing that momentum toward empathetic identification. This, in turn, could only realienate white readers fearful of manipulation and suspicious of rhetoric as the mask of an uncandid man. The adoption of the American jeremiad as the structure within which he would undertake his rhetorical experiments helped Douglass ensure a fundamental ideological consubstantiation with his white reader despite the occasional shifting designs of his rhetoric. The jeremiad identifies Douglass in a formal way; it situates him in a conventionalized relationship to his subject and his audience; its bipolar system of opposing values provides a set of standards whereby narrator and reader can negotiate their differences and achieve rapprochement. The push-pull of Douglass's rhetorical imagination, however, suspends this rapprochement at times, leaving the relationships of reader and narrator undefined, unstable, and fluid. At these key moments in the text, Douglass declares new rules and conditions whereby the white reader may approach his text and reach an understanding of and with it.

One such crucial rhetorical moment in the *Narrative* occurs as Douglass tries to describe his state of mind upon arrival in the free states. Usually this was a climactic moment in the slave narrative, something to be glorified and sentimentalized. But Douglass stresses the terror of betrayal.

> The motto which I adopted when I started from slavery was this—"Trust no man!" I saw in every white man an enemy, and in almost every colored man cause for distrust. It was a most painful situation; and, to understand it, one must needs experience it, or imagine himself in similar circumstances. Let him be a fugitive slave in a strange land—

a land given up to be the hunting-ground for slaveholders—whose inhabitants are legalized kidnappers—where he is every moment subjected to the terrible liability of being seized upon by his fellow-men, as the hideous crocodile seizes upon his prey!—I say, let him place himself in my situation—without home or friends—without money or credit—wanting shelter, and no one to give it—wanting bread, and no money to buy it,—and at the same time let him feel that he is pursued by merciless men-hunters, and in total darkness as to what to do, where to go, or where to stay,—perfectly helpless both as to the means of defence and means of escape,—in the midst of plenty, yet suffering the terrible gnawings of hunger,—in the midst of houses, yet having no home,—among fellow-men, yet feeling as if in the midst of wild beasts, whose greediness to swallow up the trembling and half-famished fugitive is only equalled by that with which the monsters of the deep swallow up the helpless fish upon which they subsist,—I say, let him be placed in this most trying situation,—the situation in which I was placed,—then, and not till then, will he fully appreciate the hardships of, and know how to sympathize with, the toil-worn and whip-scarred fugitive slave.

This cumulative sentence, with its showcase of rhetorical effects, climaxes Douglass's *Narrative* in a defiant yet directive way. Its fundamental purposes are to reclaim the concept of climax from conventional expectations and to declare the conditions under which a white reader could actually understand and appreciate the climax of a fugitive slave narrative. Douglass was aware that for many readers the point of highest interest in a slave narrative came when they finally knew how the fugitive escaped and what it felt like finally to be free. These were the ultimate facts that whites wanted to learn from such narratives; to "deprive . . . the curious of the gratification" that they expected from the revelation of such facts might jeopardize the discursive relationship that Douglass had been building with his reader. Yet Douglass did not want that relationship predicated on the assumption that whites could read slave narratives from the standpoint of the distanced, uncommitted, merely curious collector of facts and still expect to know what and who they were about. Douglass did not want to indulge his reader in a servile way; he wanted his reader to learn something about his or her responsibility to the text. For there to be a significant climax to the text, the white reader had to understand—in Ricoeur's sense of hermeneutical understanding—the ironies of Douglass's initially "painful situation" in the North. For the white reader to have the greatest possible emotional response to this moment in the text, he had to "experience it, or imagine himself in similar circumstances." Since the reader could not have the actual experience, the understanding of it was up to his or her capacity to imagine it.

In this statement Douglass, for the first time in Afro-American autobiography, declared a new and crucial role for the imagination as a mode of mediation, not distortion and deception, in black-white discourse. He was pointing toward an unprecedented answer to the central rhetorical problem of the slave narrative—how to build a bridge of sympathetic identification between the diametrical points of view of the northern white reader and the southern black fugitive. In the passage under consideration here, Douglass implies that such a bridge could not be extended from the pilings of fact set down by the black narrator. It had to be suspended from imaginative supports that connected each opposing shore of the discourse. That is, Douglass was calling for a genuine discursive relationship of equals in the slave narrative, one based on an active, flexible engagement of the white reader with the black text free from preconceived roles, instituted agendas, and programmed responses. As long as the black narrator played the suppliant role of purveyor of facts for the consumption of the preeminent reader, full appreciation and understanding of the slave narrative could not be attained. Imaginative self-projection of the reader into the text had to be the basic preparatory condition for the kind of understanding that Douglass wanted whites to derive from his story, the understanding of the individual emotional significance of the facts of a fugitive's life.

Thus Douglass repeatedly insisted of the white reader, "let him place himself in my situation" if he wished to appreciate and "know how to sympathize with" the struggle of the fugitive. This seems to have been the kind of knowledge with which Douglass hoped his *Narrative* would climax for his reader. Equally climactic is Douglass's declaration that whites had to learn how to sympathize with the fugitive slave. What accounted for their lack of the knowledge of how to do this? Perhaps Douglass was suggesting that the kind of knowledge traditionally sought from slave narratives fed only the dominant race's appetite for curious, exciting, or pathetic details of the life of the subjugated race. Such knowledge might stimulate the sentimental reflexes of the comfortable and secure toward unfortunates below them. But real sympathy, Douglass implies, could only come through an imaginative leap into the total situation of the fugitive and the world of the text.

Douglass does not talk about how the white reader could be prepared and guided in his imaginative leap into sympathetic understanding of the black narrative. His emphasis is on the reader's disposition to make that effort, and on its necessity ("then, and not till then") before real understanding could result. As we have seen, Douglass obviously understood some of the rhetorical means by which readers might be moved imaginatively to confront or distance themselves from the text. Yet he would not speak openly of the role of the black autobiographer's creative imagination in activating the white reader's sympathetic imagination. To do so would have contradicted his leader's prefatory promise that Douglass's *Narrative* contained "nothing drawn from the imagination." At this point in his literary career, Douglass, still in many ways Garrison's man, would not authorize himself at Garrison's expense. Nor would he try to authorize the slave narrative as an imaginative act at a time when its factuality and supposed eschewal of the arts of rhetoric and invention were still its greatest selling point. Nevertheless, what Douglass did declare with regard to the centrality of the imagination in the Afro-American autobiographical enterprise helped to open up the rhetorical options of black autobiographers as they had never been explored before. For if the activation of the imagination were the sine qua non for the understanding of the slave narrator's situation, then by unstated but plain logic, any rhetorical means, based more or less in fact, could be justified in an autobiography as long as it enhanced the reader's sympathetic imagination and thus his comprehension of the facts in question. Douglass set the example by appointing himself the country's black Jeremiah so as to rekindle the flame of his reader's nationalistic imagination and fan the fires of sectional division in America. After Douglass, many black autobiographers felt empowered to try out even bolder strategies of enlisting the sympathetic imagination of American readers in the cause of freedom.

Chronology

1818	Frederick Augustus Washington Bailey is born in February at Holm Hill Farm on Tuckahoe Creek, Talbot County, Maryland.
1824	Relocated at Lloyd Plantation, Wye River, at the domicile of his master, Aaron Anthony.
1825	Visited for the last time by his mother, who dies within a year.
1826	Relocated with Hugh Auld family in Fells Point area of Baltimore. Aaron Anthony dies.
1827	The slaves of Anthony's estate are divided among his heirs. Frederick is inherited by Thomas Auld, and returns to Hugh Auld's family in Baltimore. Sophia Auld begins to teach Frederick to read.
ca. 1831	Having undergone a religious conversion, Frederick joins Bethel A.M.E. Church. Purchases his first book, *The Columbian Orator*.
1832	His sister Sarah is sold to Perry Cohee of Mississippi.
1833	Frederick is moved to St. Michaels to live with Thomas Auld.
1834	Begins his year as a field hand under Edward Covey, known as a "slave-breaker," and is lashed repeatedly. Eventually fights back and is not whipped thereafter.
1835	Begins his years as field hand to William Freeland.
1836	When a plan to escape is foiled Frederick and others involved are jailed in Easton. Is returned to Baltimore by Thomas Auld.
1836–38	Works as an apprentice in the calking trade, where he is brutally beaten by white coworkers. Meets Anna Murray.

1838	He escapes north, travelling by train and boat; Anna joins him. They are married in New York City on September 15. Settles in New Bedford, Massachusetts. Takes the name Frederick Douglass.
1839	Addresses a meeting of New Bedford blacks, arguing against African colonization. Daughter Rosetta born on June 24. Later that year, he first hears the oratory of such abolitionist leaders as William Lloyd Garrison and Wendell Phillips. He is galvanized by the abolitionist creed as a "new religion."
1840	Son Lewis born on October 9.
1841	Garrison hears Douglass speak at New Bedford antislavery meeting. Douglass thrice addresses large (white) audiences at a Nantucket convention, arousing an enthusiastic response. The Massachusetts Anti-Slavery Society contracts him as a lecturer for a three-month trial period. He is ejected by force from an Eastern Railroad train for refusing to ride in the car reserved for Negroes. Moves his family from New Bedford to Lynn, Massachusetts.
1842	Is hired full-time as an antislavery orator after a 3,500 mile tour draws large and enthusiastic crowds. Son Frederick born on March 3.
1843	Douglass is attacked by a proslavery mob at Pendleton, Indiana; although his right hand is broken, he continues his lecture tour.
1844	Son Charles Remond born on October 21.
1845	*Narrative of the Life of Frederick Douglass, an American Slave* published. Douglass sets sail for Great Britain aboard Cunard steamer *Cambria,* where he is required to travel in steerage. Proslavery *Cambria* passengers threaten to throw him overboard when he attempts to speak in favor of abolition. Visits "the home of my paternal ancestors" in Liverpool. Travels to Dublin, Ireland, to begin a three-month lecture tour of several cities. Thomas Auld sells his brother Hugh title to Frederick for $100.
1846	Douglass tours Scotland and England where he is lionized by the crowds. Hugh Auld agrees to Fred-

Chronology / 185

	erick's manumission in return for £150 raised by his British boosters. When the manumission papers are filed in Baltimore, he becomes a free man.
1847	Returns to Boston. Starts newspaper, the *North Star,* in Rochester, where he settles.
1848	Attends first Women's Rights Convention at Seneca Falls, New York; joins the crusade for women's suffrage, which proves a life-long concern.
1849	Daughter Annie born on March 22.
1851	Breaks openly with Garrison over the use of political action to abolish slavery, which Garrison opposes. *North Star* is renamed *Frederick Douglass's Paper.* Receives funding from Gerrit Smith, a wealthy political activist hostile to Garrison. Becomes involved with the abolitionist Liberty Party, lead by Smith.
1855	*My Bondage and My Freedom* published.
1856	Endorses the Republican presidential candidate, John C. Fremont Smith.
1858	With John Brown, whom Douglass first met in 1848, develops ways of encouraging slave revolt.
1859	Meets Brown secretly near Chambersburg, Pennsylvania, where he learns of the plan to attack Harper's Ferry but refuses to join. Leaves Philadelphia upon hearing of the raid, and travels to Canada to evade being arrested as an accomplice. Sets sail from Quebec to England.
1860	Returns to the United States, having learned of the death of his daughter Annie.
1863	Travels through North recruiting black troops for the Union Army. Sons Lewis and Charles Remond are among the first to enlist. Visits President Lincoln, decrying discrimination against black troops. Ceases publication of *Douglass's Monthly* (successor to *Frederick Douglass's Paper*).
1864	Confers with Lincoln about the reelection campaign; decides to endorse him.
1865–66	Criticizes President Johnson's "soft" reconstruction plan, endorses Radical Republican proposals calling for black voting rights throughout the South.

1867	Sees his brother Perry for the first time since the division of Aaron Anthony's estate, and arranges for him and his family to live in Rochester.
1868	Campaigns for Ulysses S. Grant as president.
1869	Breaks with feminist leaders, who refused to support the ratification of the Fifteenth Amendment unless it includes right to vote for all women as well as black men.
1870	Joins Washington-based *New National Era* as corresponding editor; becomes editor later that year and finally buys the paper outright.
1871	Douglass is appointed assistant secretary of the commission of inquiry to Santo Domingo; later he defends Grant's plan to annex it.
1872	Although the Equal Rights Party nominates him as their vice-presidential candidate, he decides to campaign for Grant's reelection. Fire destroys Douglass's Rochester home along with many important papers. Douglass moves his family to Washington.
1874	Freedmen's Bank names Douglass as its president, but it is in financial trouble and soon fails. The *New National Era* folds.
1877	President Hayes appoints Douglass United States marshal for the District of Columbia.
1878	Acquires Cedar Hill, fifteen-acre estate in Anacostal, D.C.
1881	President Garfield appoints Douglass recorder of deeds for the District of Columbia. His third autobiography, *Life and Times of Frederick Douglass,* is published in November, but is financially a failure.
1882	Wife Anna dies after a long illness.
1884	Marries Helen Pitts, a (white) former secretary of his.
1885–87	Douglass and his new wife travel through England, France, Italy, Egypt, and Greece.
1889	President Benjamin Harrison appoints Douglass minister resident and consul general to Haiti.
1891	Resigns post, protesting attempts by the U.S. State Department and some American businessmen to acquire Mole St. Nicolas.

1892–93	Serves as commissioner of Haitian exhibit at World's Fair in Chicago. Duped by dishonest promoters, he announces his plan to establish Freedom Manufacturing Co., a textiles concern, near Norfolk, Virginia, expecting to employ several hundred blacks. The project turns out to be a fraud, with the promoters hoping to trade on the prestige of Douglass's name.
1894	Gives his final great speech, "Lessons of the Hour," assailing the lynch law in the South.
1895	Dies of heart failure at Cedar Hill, February 20, and is buried in Mount Hope Cemetery, Rochester.

Contributors

HAROLD BLOOM, Sterling Professor of the Humanities at Yale University, is the author of *The Anxiety of Influence, Poetry and Repression,* and many other volumes of literary criticism. His forthcoming study, *Freud: Transference and Authority,* attempts a full-scale reading of all of Freud's major writings. A MacArthur Prize Fellow, he is general editor of five series of literary criticism published by Chelsea House. During 1987–88, he served as Charles Eliot Norton Professor of Poetry at Harvard University.

ALBERT E. STONE is Professor of American Studies at the University of Iowa. He is the author of *Autobiographical Occasions and Original Acts: Versions of American Identity from Henry Adams to Nat Shaw* and *Innocent Eye: Childhood in Mark Twain's Imagination.*

H. BRUCE FRANKLIN is Professor of English and American Literature at Rutgers University, Newark College of Arts and Sciences. His books include *Future Perfect* and *Back Where You Came From.*

ROBERT B. STEPTO is Professor of English, American Studies, and African and Afro-American Studies at Yale University and the author of *Behind the Veil: A Study of Afro-American Narrative.*

HENRY LOUIS GATES, JR., is a Professor of English, Comparative Literature, and Africana Studies at Cornell University. A MacArthur Prize Fellow, he is the author of *Figures in Black* and *The Signifying Monkey.*

ROBERT G. O'MEALLY teaches English and Afro-American studies at Wesleyan University. He is the author of *The Craft of Ralph Ellison.*

HOUSTON A. BAKER, JR., is the Albert M. Greenfield Professor of Human Relations at the University of Pennsylvania. His books include *The Journey Back: Issues in Black Literature and Criticism, Long Black Song: Essays in Black American Literature and Culture,* and *Blues, Ideology and Afro-American Literature.*

ANNETTE NIEMTZOW, who formerly taught American literature, now manages the tennis courts in Central Park. She holds a Ph.D. in English from Harvard University.

ANN KIBBEY is Associate Professor of English at the University of Washington, Seattle, and the author of a study of rhetoric, prejudice, and violence in Puritan America.

JOHN SEKORA is Professor of English at North Carolina Central University. He is the author of *Luxury: The Concept in Western Thought, Eden to Smollett* and the editor, with Darwin T. Turner, of *The Art of the Slave Narrative.*

WILLIAM L. ANDREWS is Professor of English at the University of Wisconsin, Madison. He is the author of *To Tell a Free Story* and *The Literary Career of Charles W. Chesnutt,* and the editor of *Critical Essays on W. E. B. DuBois* and *Literary Romanticism in America.*

Bibliography

Baker, Houston A., Jr., *Blues, Ideology, and Afro-American Literature*. Chicago: University of Chicago Press, 1984.
———. *The Journey Back: Issues in Black Literature and Criticism*. Chicago: University of Chicago Press, 1980.
———. *Long Black Song: Essays in Black American Literature and Culture*. Charlottesville: The University Press of Virginia, 1972.
———. "The Problem of Being: Some Reflections on Black Autobiography." *Obsidian* 1, no. 1 (1974): 18–30.
Bontemps, Arna. *Great Slave Narratives*. Boston: Beacon Press, 1969.
Brawley, Benjamin, ed. *Early Negro American Writers*. Chapel Hill: University of North Carolina Press, 1969.
Butterfield, Stephen. *Black Autobiography in America*. Amherst: University of Massachusetts Press, 1974.
Clasby, Nancy T. "Frederick Douglass's *Narrative*: A Content Analysis." *CLA Journal* 14 (1971): 242–50.
Couser, G. Thomas. *American Autobiography: The Prophetic Mode*. Amherst: University of Massachusetts Press, 1979.
De Pietro, Thomas. "Vision and Revision in the Autobiographies of Frederick Douglass." *CLA Journal* 26 (1983): 384–96.
Felgar, Robert. "The Rediscovery of Frederick Douglass." *Mississippi Quarterly: The Journal of Southern Culture* 35 (1982): 427–38.
Foner, Philip S. *Frederick Douglass: A Biography*. New York: Citadel, 1964.
Gibson, Donald B. "Reconciling Public and Private in Frederick Douglass's *Narrative*." *American Literature* 57 (1985): 549–69.
Haskett, Norman D. "Afro-American Images of Africa: Four *Antebellum* Black Authors." *Ufahamu* 3, no. 1 (1972): 29–40.
Jugurtha, Lillie Butler. "Point of View in the Afro-American Slave Narratives: A Study of Narratives by Douglass and Pennington." In *The Art of the Slave Narrative*, edited by John Sekora and Darwin T. Turner. Macomb: Western Illinois University: Essays in Literature Books, 1982.
Loggins, Vernon. *The Negro Author: His Development in America*. New York: Columbia University Press, 1931.
MacKethan, Lucinda H. "Metaphors of Mastery in the Slave Narratives." In *Art of*

the Slave Narrative: Original Essays in Criticism and Theory, edited by John Sekora and Darwin T. Turner. Macomb: Western Illinois University, 1982.

Matlock, James. "The Autobiography of Frederick Douglass." *Phylon* 40 (March 1979): 15–28.

Nichols, Charles Harold. *Many Thousands Gone: The Ex-Slaves' Account of Their Bondage and Freedom.* Leiden: Brill, 1963.

Nichols, William W. "Individualism and Autobiographical Art: Frederick Douglass and Henry Thoreau." *CLA Journal* 16 (1972): 145–58.

Olney, James. " 'I Was Born': Slave Narratives, Their Status as Autobiography and as Literature." In *The Slave's Narrative,* edited by Charles T. Davis and Henry Louis Gates, Jr. New York: Oxford University Press, 1985.

Ostendorf, Bernd. "Violence and Freedom: The Covey Episode in Frederick Douglass's Autobiography." In *Mythos und Aulklarung in der amerikanischen Literatur/ Myth and Enlightenment in American Literature,* edited by Dieter Meindl and Friedrich W. Horlacher. Erlangen: Universitatsbund Erlangen-Nurberg, 1985.

Piper, Henry Dan. "The Place of Frederick Douglass's Narrative of the Life of an American Slave in the Development of a Native American Prose Style." *The Journal of Afro-American Issues* 5, no. 2 (1977): 183–91.

Quarles, Benjamin. *Frederick Douglass.* Washington: Associated Publishers, 1968.

———. "Frederick Douglass: Black Imperishable." *Quarterly Journal of the Library of Congress* 29 (1972): 159–69.

———. Introduction to John Harvard Library edition of the *Narrative of the Life of Frederick Douglass.* Cambridge: Harvard University Press, 1960.

———. "Narrative of the Life of Frederick Douglass." In *Landmarks of American Writing,* edited by Hennig Cohen. New York: Basic Books, 1969.

Ripley, Peter. "The Autobiographical Writings of Frederick Douglass." *Southern Studies* 24 (1985): 5–29.

Sekora, John. "The Dilemma of Frederick Douglass: The Slave Narrative as Literary Institution." *Essays in Literature* 10, no. 2 (Fall 1983): 219–26.

Stepto, Robert B. *Behind the Veil: A Study of Afro-American Narrative.* Urbana: University of Illinois Press, 1979.

Sundquist, Eric J. "Frederick Douglass: Literacy and Paternalism." *Raritan* 6, no. 2 (Fall 1986): 108–24.

Takaki, Ronald T. "Not Afraid to Die: Frederick Douglass and Violence." In *Violence in the Black Imagination,* 17–35. New York: Putnam, 1972.

Terry, Eugene. "Black Autobiography: Discernible Forms." *Okike* 19 (September 1981): 6–10.

Walker, Peter F. *Moral Choices: Memory, Desire, and Imagination in Nineteenth Century American Abolition.* Baton Rouge: Louisiana University Press, 1978.

Yellin, Jean Fagan. *The Intricate Knot: Black Figures in American Literature, 1776– 1863.* New York: New York University Press, 1972.

Zeitz, Lisa Margaret. "Biblical Allusion and Imagery in Frederick Douglass's *Narrative.*" *CLA Journal* 25 (1981): 56–64.

Acknowledgments

"Identity and Art in Frederick Douglass's *Narrative*" by Albert E. Stone from *CLA Journal* 17, no. 2 (December 1973), © 1973 by the College Language Association. Reprinted by permission of the College Language Association.

"Animal Farm Unbound" (originally entitled "Animal Farm Unbound, Or, What the *Narrative of the Life of Frederick Douglass, an American Slave* Reveals About American Literature") by M. Bruce Franklin from *New Letters* 43, no. 3 (April 1977), © 1977 by the Curators of the University of Missouri, Kansas City. Reprinted by permission of the author, *New Letters,* and the Curators of the University of Missouri, Kansas City.

"Narration, Authentication, and Authorial Control" (originally entitled "Narration, Authentication, and Authorial Control in Frederick Douglass' *Narrative* of 1845") by Robert B. Stepto from *Afmerican Literature: The Reconstruction of Instruction,* edited by Dexter Fisher and Robert B. Stepto, © 1978 by the Modern Language Association of America. Reprinted by permission of the Modern Language Association of America.

"Binary Oppositions in Chapter One of the *Narrative*" (originally entitled "Binary Oppositions in Chapter One of *Narrative of the Life of Frederick Douglass, an American Slave, Written by Himself*") by Henry Louis Gates, Jr., from *Afro-American Literature: The Reconstruction of Instruction,* edited by Dexter Fisher and Robert B. Stepto, © 1978 by the Modern Language Association of America. Reprinted by permission of the Modern Language Association of America.

"The Text Was Meant to Be Preached" (originally entitled "Frederick Douglass' 1845") by Robert G. O'Meally from *Afro-American Literature: The Reconstruction of Instruction,* edited by Dexter Fisher and Robert B. Stepto, © 1978 by the Modern Language Association of America. Reprinted by permission of the Modern Language Association of America.

"Autobiographical Acts and the Voice of the Southern Slave" by Houston A. Baker, Jr., from *The Journey Back: Issues in Black Literature and Criticism* by Houston A. Baker, Jr., © 1975 by Houston A. Baker, Jr. Reprinted by permission of the author. This essay originally appeared in *Obsidian: Black Literature in Review* 1, no. 1 (1975), © 1975 by Houston A. Baker, Jr. Reprinted by permission of the author and *Obsidian: Black Literature in Review.*

"The Problematic of Self in Autobiography: The Example of the Slave Narrative" by Annette Niemtzow from *The Art of Slave Narrative: Original Essays in Criticism and Theory,* edited by John Sekora and Darwin D. Turner, © 1982 by Western Illinois University. Reprinted by permission.

"Language in Slavery" (originally entitled "Language in Slavery: Frederick Douglass' *Narrative*") by Ann Kibbey from *Prospects: The Annual of American Cultural Studies* 8 (1983), © 1983 by Cambridge University Press. Reprinted by permission of Cambridge University Press.

"Comprehending Slavery: Language and Personal History in the *Narrative*" (originally entitled "Comprehending Slavery: Language and Personal History in Douglass' *Narrative* of 1845") by John Sekora from *CLA Journal* 29, no. 2 (December 1985), © 1985 by the College Language Association. Reprinted by permission of the College Language Association.

"The Performance of the *Narrative*" (originally entitled "The Performance of Slave Narrative") by William L. Andrews from *To Tell a Free Story* by William L. Andrews, © 1986 by the Board of Trustees of the University of Illinois. Reprinted by permission of the University of Illinois.

Index

Adams, Henry: Douglass compared to, 11; *The Education of Henry Adams,* 97
American black literature: the black sermon's influence on, 80–81; slave narrative's influence on, 6, 16, 27, 47, 63
Angelou, Maya, *I Know Why the Caged Bird Sings,* 27
Animals, Douglass and imagery of, 29–30, 32–34, 36–37, 38–40, 42, 67–69, 74, 99, 119–20, 143
Anthony, Aaron, 89–90, 157, 158, 175, 177, 178
Auld, Mr. and Mrs. Hugh and Sophie: Douglass and, 3–4, 18, 20, 34–35, 52, 86, 91, 100–101, 103, 106, 109–10, 133, 143, 156–57, 159, 167, 170–71, 175
Autobiographical focus: of the *Narrative,* 9–10, 12–16, 17–18, 21, 53, 56, 57, 67–68, 69–70, 181–82
Autobiography: Butterfield on, 60; Cooke on, 74; Croce on, 95; Gusdorf on, 115; Hart on, 10; nature of, 10, 60–61, 64–65, 95, 96–98, 114–15, 118, 120, 122, 124, 127, 129–30, 153; Wordsworth on, 114, 120
Autobiography (Franklin), 64, 118–19, 122, 126
Autobiography of an Ex-Coloured Man (Johnson), 55, 63
Autobiography of Malcolm X, The, 8, 27; Handler's introduction to, 13

Baker, Houston A., Jr., 10; *Long Black Song,* 11, 29; on the *Narrative,* 11; on nature of slavery, 116–17
Baldwin, James: black sermon's influence on, 80–81; and Christianity, 93; compared to Douglass, 6
Ball, Charles, *Fifty Years in Chains,* 10
Bercovitch, Sacvan, on the jeremiad, 166, 168
Bibb, Henry, 21, 150
Binary opposition in language: Jakobson on, 66; Jameson on, 67; Levi-Strauss on, 66; of the *Narrative,* 71–73, 173, 174–75
Bingham, Caleb, *The Columbian Orator,* 18
Bird, Robert Montgomery, 171
Black Boy (Wright), 27, 55; Fisher's introduction to, 13
Black Image in the White Mind, The (Frederickson), 30
Black inferiority, "scientific" proof of, 30–31
Black literature, American. See American black literature
Black Man: His Antecedents, His Genius, and His Achievements, The (Brown), 7–8
Black sermon: influence on American black literature, 80–81; influence on James Baldwin, 80–81; Johnson on, 79; structure and style of, 78–80, 94
Bonifacius, An Essay Upon the Good (Mather), 118–19
Bontemps, Arna, 10
Brawley, Benjamin, 10
Brent, Linda, *Incidents in the Life of a Slave Girl,* 124–26, 127–28, 129
Brown, Claude: compared to Douglass, 16; *Manchild in the Promised Land,* 27
Brown, Henry "Box", slave narrative of, 55, 160
Brown, William Wells: on the *Narrative,* 8; novels by, 9; slave narrative of, 7, 128, 160; *The Black Man: His Antecedents, His Genius, and His Achievements,* 7–8
Bunyan, John, *The Pilgrim's Progress,* 150–51
Butterfield, Stephen: on autobiography, 60; on slave narrative, 65–66

Child, Lydia Maria, 10, 61
Christianity: Baldwin and, 93; Douglass and, 4–6, 57, 72, 77, 92, 93, 103–4, 105, 147–48, 149, 150, 166
Chronotype, on slave narrative, 25

195

Clarissa Harlowe (Richardson), 126
Clarke, Lewis and Milton, slave narrative of, 7
Claseby, Nancy, on Douglass, 41–42
Coffin, William C., 14, 108, 132
Colfax, Richard: *Evidence Against the Views of the Abolitionists, Consisting of Physical and Moral Proofs of the Natural Inferiority of the Negroes,* 30–31, 37
Columbian Orator, The (Bingham), Douglass on, 18, 101, 169
Cooke, Michael, on autobiography, 74
Cooper, James Fenimore, 171
Covey, Edward: description of, 23, 90, 140; Douglass and, 3, 4, 11, 12, 13, 22, 26, 37–38, 39, 53, 87–88, 123, 160, 168–69, 175
Craft, William and Ellen: slave narrative of, 55, 150–51, 160
Crania Aegyptiaca (Morton), 31
Crania Americana (Morton), 31
Critical reception, of *Narrative of the Life of Frederick Douglass,* 62–63
Croce, Benedetto, on autobiography, 95
Curry, James, slave narrative of, 159

Dialogue Between a Master and a Slave (Aiken), 19
Diaspora, slavery as, 97–98
Discourse on Manumission of Slaves (Miller), 18
Douglass, Frederick: adoption of name by, 18, 121–22, 171; and Christianity, 4–6, 57, 72, 77, 92, 93, 103–4, 105, 147–48, 149, 150, 166; Claseby on, 41–42; compared to Henry Adams, 11; compared to James Baldwin, 6; compared to Claude Brown, 16; compared to Benjamin Franklin, 11, 16, 118–19, 169; compared to Malcolm X, 16; compared to Mather, 11; compared to Melville, 43; and Edward Covey, 3, 4, 11, 12, 13, 22, 26, 37–38, 39, 53, 87–88, 123, 160, 168, 169, 175; escape of, 17, 20, 21, 42, 54, 56, 104, 151, 161, 168, 171, 180; ethical attitudes of, 92; Fanon on, 41; Garrison as sponsor and guarantor of, 13–14, 15, 21, 48–50, 57, 81–82, 106–7, 132, 133, 154–55, 156, 160, 161, 178, 182; hired out as craftsman, 22–23, 170–71; on his Aunt Hester, 3, 15, 32–33, 72–73, 89–90, 137–38, 139, 167, 175–76; on his parents, 1–3, 6, 32, 70–71, 99, 117, 118–19, 136–37, 138; and imagery of animals, 29–30, 32–34, 36–37, 38–40, 42, 67–69, 74, 99, 119–20, 143; and imagery of ships, 20–21, 24, 26, 29, 102, 110, 149, 160; and Lincoln's second inaugural reception, 26; and literacy, 18–20, 23, 34–36, 52, 86–87, 100, 101–2, 105–6, 114–15, 121, 159, 167; married to Anna Murray, 104, 161, 171; as Marshal of District of Columbia, 109, 110; and Mr. and Mrs. Auld, 3–4, 18, 20, 34–35, 52, 86, 91, 100–101, 103, 106, 109–10, 133, 143, 156–57, 159, 167, 170–71, 175; *My Bondage and My Freedom,* 9, 12, 25, 154, 162; as Old Testament hero, 92–93; on Patrick Henry, 49, 170; Phillips as sponsor and guarantor of, 13–14, 15, 48, 50–51, 81, 82, 109, 154, 160, 161; on racism in the North, 172; and Reconstruction, 26; on scriptural excuses for racism, 72, 83; on slaves' songs, 17, 53–54, 74–75, 84, 91, 134–36, 158–59, 175; on *The Columbian Orator,* 18, 101, 169; *The Life and Times of Frederick Douglass,* 12, 21, 25–26, 109–10, 172

Economic and Philosophic Manuscripts of 1844 (Marx), 31
Education of Henry Adams, The (Adams), 97
Edwards, Jonathan: *Personal Narrative,* 96, 97; Sayre on, 96
Ellison, Ralph, *The Invisible Man,* 130
Everett, Edward, *Slaves in Barbary: A Drama in Two Acts,* 18
Evidence Against the Views of the Abolitionists, Consisting of Physical and Moral Proofs of the Natural Inferiority of the Negroes (Colfax), 30–31, 37

Fanon, Frantz, on Douglass, 41
Fifty Years in Chains (Ball), 10
Franklin, Benjamin: *Autobiography,* 64, 118–19, 122, 126; Douglass compared to, 11, 16, 118–19, 169
Franklin, H. Bruce: "On the Teaching of Literature in the Highest Academies of the Empire," 40–41
Frederickson, George, *The Black Image in the White Mind,* 30
Freeland, William, 22, 26, 52, 104, 157, 168
Freud, Sigmund, and interpretation of the *Narrative,* 4, 6
Fugitive Blacksmith, The (Pennington), 161, 162–63
Fundamentals of Language (Jakobson and Halle), 66

Garnet, Rev. Henry Highland, 93
Garrison, William Lloyd: motives of, 49–50; as sponsor and guarantor of Douglass, 13–14, 15, 21, 48–50, 56, 57, 81–82, 106–7, 132, 133, 154–55, 156, 160, 161, 178, 182

Gore, Austin, description of, 3, 4, 140–43, 157, 175
Graham, George R., on slave narrative, 62
Grandy, Moses, slave narrative of, 159, 165, 170
Gulliver's Travels (Swift), 39
Gusdorf, George, on autobiography, 115

Hammon, Jupiter, 104, 109, 153, 154
Hart, F. R., on autobiography, 10
Hawthorne, Nathaniel: "My Kinsman, Major Molineux," 34; preface to *The Marble Faun*, 96
Henry, Patrick, Douglass on, 49, 170
Henson, Josiah, *Truth Stranger than Fiction*, 7, 9, 26, 165, 169, 170
Hildreth, Richard, *The Slave: or Memoirs of Archy Moore*, 9
History, the *Narrative* as, 9–10
Hopkins, Rev. Rigby, description of, 176–77, 178

I Know Why the Caged Bird Sings (Angelou), 27
Imagery: in the *Narrative*, 15–16, 21–23, 26
Incidents in the Life of a Slave Girl (Brent), 124–26, 127–28, 129
Incidents in the Life of a Slave Girl (Jacobs), 10, 61, 154
Intricate Knot, The (Yellin), 10–11
Invisible Man, The (Ellison), 130

Jacobs, Harriet, 21; *Incidents in the Life of a Slave Girl*, 10, 61, 154
Jakobson, Roman, on binary opposition in language, 66
Jakobson, Roman and Morris Halle, *Fundamentals of Language*, 66
James, Henry, 116
Jameson, Frederic: on binary opposition in language, 67; *The Prison-House of Language*, 67
Jenkins, Sandy [slave], 22, 77, 87–88, 92, 123
Jeremiad: Bercovitch on the, 166, 168; the *Narrative* as, 165–68, 173–74, 182
Jews in captivity, slavery compared to, 91
Johnson, James Weldon: *Autobiography of an Ex-Coloured Man*, 55, 63; on black sermon, 79

Kemble, Fanny, and slaves' songs, 17

Lamming, George, 103
Lane, Lunsford, slave narrative of, 124, 159, 169, 170
"Learning Theory" (McConnell), 39
Leaves of Grass (Whitman), 97
Levi-Strauss, Claude, on binary opposition in language, 66
Life and Times of Frederick Douglass, The, 12, 21, 25–26, 109–10, 172; Logan on, 26
Life of John Thompson, The (Thompson), 65–66
Lincoln, Abraham, Douglass and his second inaugural reception, 26
Literacy, Douglass and, 18–20, 23, 34–36, 52, 86–87, 100, 101–2, 105–6, 114–15, 121, 159, 167
Literature, American black. *See* American black literature
Lloyd, Edward, 17, 83, 90, 91, 110, 119, 133, 134, 139–40, 146, 157, 158, 159, 167
Logan, Rayford, on *The Life and Times of Frederick Douglass*, 26
Loggins, Vernon, 10
Long Black Song (Baker), 11, 29

McConnell, James, "Learning Theory," 39
Magnalia Christi Americana (Mather), 118
Malcolm X: compared to Douglass, 16; *The Autobiography of Malcolm X*, 8, 13, 27
Manchild in the Promised Land (Brown), 27
Many Thousand Gone (Nichols), 8
Marble Faun, The (Hawthorne), preface to, 96
Marx, Karl: *Economic and Philosophic Manuscripts of 1844*, 31; and slavery, 145, 146–47
Mather, Cotton: *Bonifacius, An Essay Upon the Good*, 118–19; Douglass compared to, 11; *Magnalia Christi Americana*, 118
Melville, Herman: as abolitionist, 165; Douglass compared to, 43; literary style of, 65–66; *Moby-Dick*, 65–66; *Pierre; Or, The Ambiguities*, 34, 64; *Typee*, 43
Micromegas (Voltaire), 39
Miller, Samuel, *Discourse on Manumission of Slaves*, 18
Moby-Dick (Melville), 65–66
Morton, Samuel George: *Crania Aegyptiaca*, 31; *Crania Americana*, 31
Moses, Wilson J., on the slave narrative, 165
Murray, Anna, Douglass's marriage to, 104, 161, 171
My Bondage and My Freedom (Douglass), 9, 12, 25, 154, 162
"My Kinsman, Major Molineux" (Hawthorne), 34

Nabokov, Vladimir, *Speak, Memory*, 21
Nantucket Convention of *1841*, 13–14, 21, 49,

Nantucket Convention of *1841* (*continued*)
50, 56, 106–7, 108, 131, 132–33, 135, 147, 149, 162, 168
Narrative of James Williams, 12
Narrative of the Life of Frederick Douglass: accused of fraud, 12–13, 156; autobiographical focus of, 9–10, 12–16, 17–18, 21, 53, 56, 57, 67–68, 69–70, 181–82; Baker on, 11; binary opposition of language in, 71–73, 173, 174–75; Brown on, 8; critical reception of, 62–63; cultural importance of, 8–9; Freudian interpretation of, 4, 6; as history, 9–10; imagery in, 15–16, 21–23, 26; as jeremiad, 165–68, 173–74, 182; Ephraim Peabody on, 7, 10, 175, 179; as polemic, 11; popularity of, 8, 62; Quarles on, 11; as sermon, 78, 81, 82–83, 85, 86, 87, 89, 91, 92–94; as spiritual autobiography, 11, 167; structure and style of, 9, 11, 23–24, 32–33, 48–49, 51–53, 54, 56–57, 67–69, 70–71, 73–75, 85, 86–87, 89, 91–92, 93–94, 99–101, 102–5, 107, 109, 131–32, 133–34, 136, 138, 140–41, 146–47, 151–52, 167–68, 169, 171, 173–75, 176–78, 179–82; use of parables in, 86–88; verisimilitude of, 9; Yellin on, 10–11, 26
Nichols, Charles H., 10; *Many Thousand Gone*, 8; on slave narrative, 8, 25
Northup, Solomon, 16, 21, 131

Old Testament hero, Douglass as, 92–93
"On the Teaching of Literature in the Highest Academies of the Empire" (Franklin), 40–41

Pamela (Richardson), 126
Parables, use of in the *Narrative*, 86–88
Peabody, Ephraim: on the *Narrative*, 7, 10, 175, 179; on slave narrative, 7–8
Pennington, James W. C., 16, 149–50, 161; *The Fugitive Blacksmith*, 161, 162–63
Personal Narrative (Edwards), 96, 97
Phillips, Ulrich B., on slave narrative, 115
Phillips, Wendell: on racism in the North, 173; as sponsor and guarantor of Douglass, 13–14, 15, 81, 82, 109, 154, 160, 161
Picaresque literature, slave narrative compared to, 60–61, 63
Pierre; Or, The Ambiguities (Melville), 34, 64
Pilgrim's Progress, The (Bunyan), 150–51
Planet of the Apes [film], 39
"Plantation novel," popularity of, 61, 83
Polemic, the *Narrative* as, 11
Prison-House of Language, The (Jameson), 67
Protestantism, and slavery, 3, 4–5

Quarles, Benjamin, 10; on the *Narrative*, 11

Racism, Douglass on scriptural excuses for, 72, 83
Racism in the North: Douglass on, 172; Phillips on, 173
Rebellion, distinguished from revolution, 166–67
Reconstruction, Douglass and, 26
Richardson, Samuel: *Clarissa Harlowe*, 126; *Pamela*, 126
Roper, Moses, slave narrative of, 159, 160

Sadism, slaveholding and, 1–2, 3–4, 6
Sayre, Robert, on Jonathan Edwards, 96
"Scientific" proof, of black inferiority, 30–31
Sermon, the *Narrative* as, 78, 81, 82–83, 85, 86, 87, 89, 91, 92–94
Sermon, black. *See* Black sermon
Severe, Mr., 3, 4, 22, 52, 90, 140, 157
Sexual imagery, in the slave narrative, 32–33, 127–29
Ships, Douglass and imagery of, 24, 26, 29, 29–31, 102, 110, 149, 160
Simms, William Gilmore, 171
Slave: or Memoirs of Archy Moore, The (Hildreth), 9
Slave market, 143–45
Slave narrative: authentication of, 13–14, 15, 21, 45–48, 49–51, 55, 57; Butterfield on, 65–66; *Chronotype* on, 25; compared to picaresque literature, 60–61, 63; generic form of, 46–48; Graham on, 62; influence on American black literature, 6, 16, 27, 47, 63; Moses on, 165; nature of, 45–48, 65, 113–14, 115–16, 123, 124–25, 128–29, 130, 153–55, 161–63; Nichols on, 8, 25; Ephraim Peabody on the, 7–8; Phillips on, 115; political effect of, 25; popularity of, 8, 60–62; sexual imagery in, 32–33, 127–29; Starling on popularity of, 62; structure and style of, 45–48; Stuckey on, 98
Slaveholding: psychology of, 1, 3–4, 6; and sadism, 1–2, 3–4, 6
Slavery: chattel principles of, 29, 36, 98, 116–17, 136, 140–42, 143–45, 146–47, 156; compared to the Jews in captivity, 91; as diaspora, 97–98; economics of, 3; and expansion of agricultural economy, 30; Marx and, 145, 146–47; Protestantism and, 3, 4–5
Slaves in Barbary: A Drama in Two Acts (Everett), 18

Slaves' songs, Douglass on, 17, 53–54, 74–75, 84, 91, 134–36, 158–59, 175
Speak, Memory (Nabokov), 21
Starling, Marion Wilson, on popularity of slave narrative, 62
Stowe, Harriet Beecher, 9; *Uncle Tom's Cabin*, 8, 61
Stuckey, Sterling, on slave narrative, 98
Swift, Jonathan, *Gulliver's Travels*, 39

Thompson, John: literary style of, 65–66; slave narrative of, 160; *The Life of John Thompson*, 65–66
Thoreau, Henry David, as abolitionist, 165
Truth Stranger than Fiction (Henson), 7, 9, 26, 165, 169, 170
Tucker, George, *The Valley of the Shenandoah*, 61
Typee (Melville), 43

Uncle Tom's Cabin (Stowe), 8, 61

Valley of the Shenandoah, The (Tucker), 61
Vassa, Gustavus, slave narrative of, 98, 103, 109, 110, 149
Voltaire, *Micromegas*, 39

Washington, Booker T., 111
Watson, Henry, slave narrative of, 7
Wheatley, Phyllis, 98, 109
Whitman, Walt, *Leaves of Grass*, 97
Wordsworth, William, on autobiography, 114, 120
Wright, Richard, *Black Boy*, 13, 27, 55

Yellin, Jean Fagan: *The Intricate Knot*, 10–11; on the *Narrative*, 10–11, 26